Materials Matter

Seok-Woo Lee & Wyeth Haddock

Designed by Seok-Woo Lee & Wyeth Haddock
Illustrations by Seok-Woo Lee
Book cover design by Kyunga Lee

ISBN 979-8-9914383-1-5

Seok-Woo Lee dedicates this book to **Kyunga Lee**, whose unwavering love and support have accompanied him throughout his journey as a materials scientist, and to **Prof. William D. Nix**, who has been an exemplary role model and has shared with him the wonder of materials science.

Wyeth Haddock dedicates this book to his amazing parents, **Tim Haddock** and **Katie Haddock**, whose constant encouragement and support made his path to becoming a materials scientist possible.

Contents

Heat treatment beyond metals
The goal of heat treatment

Preface:

The development of new materials is becoming increasingly important.

To make something, you need materials. Just like you need good ingredients to make a delicious dish, you need good materials to make a well performing product. For example, a knife made of metal is sharper and much more durable than a knife made of stone. A light-emitting-diode bulb made using a semiconductor is brighter than an incandescent light bulb made of tungsten wire and is more energy efficient (meaning that it can be used for much longer while the same amount of energy is consumed). The development of advanced materials like this benefits our lives and has a significant impact on the development of human civilization. As we all know, human history is often divided according to the main materials used in that era (Stone Age, Bronze Age, Iron Age) (**Figure 1**). The era we currently live in is sometimes referred to as the era of silicon, named after silicon, the primary material used in the semiconductor chips found in computers and cell phones. What new materials will make our world better in the future?

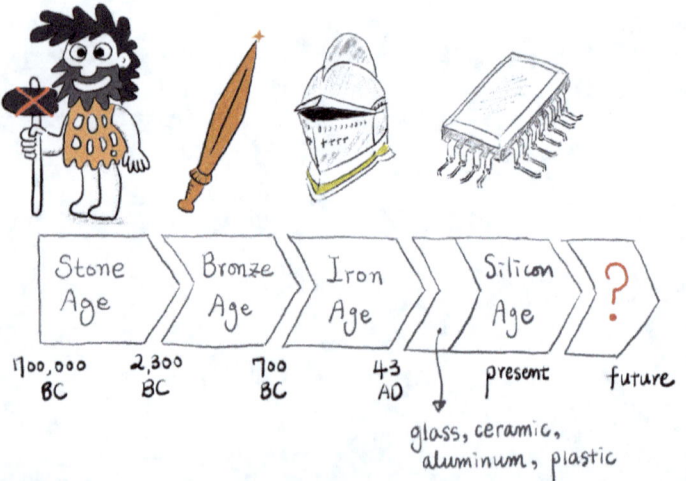

Figure 1. The history of human civilization can be divided by the materials most used during that era. In other words, the revolution in materials technology is closely related to the revolution in human civilization, so the development of new materials has a great impact on our lives.

2

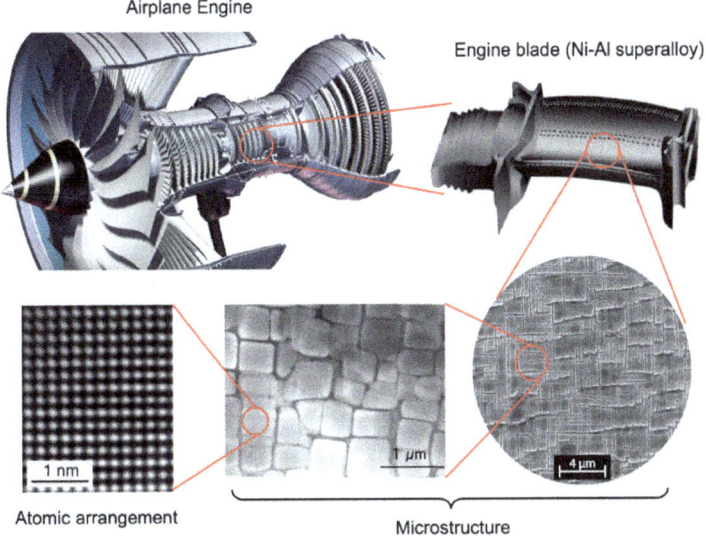

Airplane Engine

Engine blade (Ni-Al superalloy)

1 nm

1 µm

4 µm

Atomic arrangement

Microstructure

Figure 2. Airplane engine blades must be made of materials that can withstand extreme environments with temperatures above 1000 degrees Celsius for several years. This special material (nickel-aluminum superalloys) has a unique microstructure (third and fourth photos) and atomic arrangement (last photo) and can be made only by humans through a method called single crystal growth.[1-4]

In the past, natural materials were used to make things with little to no modification. Large stones could be cut and ground to make pyramids, trees could be cut down to make ships, and animal hair could be plucked, or its skin made into clothes. However, as civilization developed, a demand for high performance led to a need for new and improved materials—materials that are no longer available in nature. Therefore, humans must develop and create them themselves. The nickel-aluminum superalloy, a special material used in aircraft engine blades, is light and strong and can withstand temperatures above 1000 °C for several years, but cannot be obtained directly from nature (**Figure 2**). Transistors, the most important component of computer integrated circuits, can be made by mixing different elements in very precise amounts on a silicon wafer, but cannot be obtained directly from nature. From the Wright brothers' airplane made of wood and fabric, to Space X's rockets, many important inventions could never have been achieved without the development of new materials.

Since the development of new materials plays a central role in technological development, materials science—the discipline that studies materials development—is of indescribable importance to us. As materials scientists, we have met many young middle and high school students over the years to promote materials science. Through conversations with students and parents during departmental information sessions, we learned that many people are not familiar with materials science. They told us that they had an intuitive understanding of other engineering fields (mechanical engineering, electronic engineering, civil engineering, architectural engineering, computer engineering, etc.), but that materials science was new to them.

This led us to ask a question: why are so many people unfamiliar with materials science compared to other engineering fields? In our opinion, the first reason is that it is not easy to picture what materials scientists do. It is very easy to imagine mechanical engineers making cars, and civil engineers making buildings. Unlike cars or buildings, materials do not have a defined shape, making them harder to conceptualize. The second reason could be that materials science is not as easily accessible through the media as other engineering fields. People get a lot of information from the media. Although many of us have seen videos showing how products are assembled, we rarely learn how to obtain and make the materials that comprise the parts. We've seen a lot of video clips

Figure 3. Tesla's Giga-factory. We have seen many videos of car manufacturing in the media, but we don't remember seeing much of the process of making the materials needed for car.[5]

on the robots in car factories assembling cars quickly and accurately (**Figure 3**) but have never seen how materials scientists find and create the optimal materials necessary to make those parts.

When we were young, we watched Star Wars and dreamed of building the various spaceships seen in the movie, sparking an interest in becoming a scientist or engineer. Seok-Woo remembers being so shocked when he saw the liquid metal man in Terminator 2. After watching the movie, he had a nightmare of a liquid metal man chasing him for a week straight. Nonetheless, he wondered if it might be possible to create that strange liquid metal. Science fiction movies show cutting-edge weapons, airplanes, and robots. They are excellent audio-visual materials that stimulate interest in science and engineering but unfortunately, don't typically describe the materials science behind the scenes (**Figure 4**).

Figure 4. (Top left) There is a scene showing Iron Man's armor being assembled, but it is not said how such strong armor materials were obtained (**Paramount Pictures**). (Above right) It doesn't tell us how Superman's suit can be so strong and tough (**Warner Bros Pictures**). (Bottom left) How can we make a liquid metal robot that can be restored to its original shape even after it is seriously damaged? (**TriStar Pictures**) (Bottom right) How can we make a material that is as elastic and tough as Spiderman's spider web? (**Sony Pictures**) [6-9]

The 2008 film Iron Man shows how Tony assembles the armor but doesn't tell us in detail where and how he got the materials to make it—materials that are light and yet, ridiculously strong. Thankfully, the movie mentions that the material is a combination of gold and titanium, but from our literature survey, it doesn't seem easy to combine the two elements to create such a strong and durable material. We can make a Halloween costume that looks exactly like Superman's, but how can we make the strong, bulletproof flame-resistant material seen in the movies? What kind of material is the spider silk that Spider-Man uses, and how is it so elastic and tough? In the movie Avatar, humans invade the place where the Na'vi tribe lives to take over a stone floating in the air, but we

6

don't see what the strange stone is or how it works. We realized that the difference between movies and reality is largely determined by whether these ridiculous materials are present—almost all movies begin with the assumption that they are. Science fiction movies never tell us the mystery of how the main characters obtained such fantastic materials (because no one knows!).

To summarize, in materials science, the shape of the final product is not set, making it difficult to intuitively imagine what it looks like. Additionally, there are very few opportunities to encounter it in media or science fiction movies. To us, this is why young students are not familiar with materials science (of course, this is our opinion).

Throughout Seok-Woo's career as a materials scientist, he has noticed that the creation of new materials has become increasingly important in technological development. In our grandparents' and

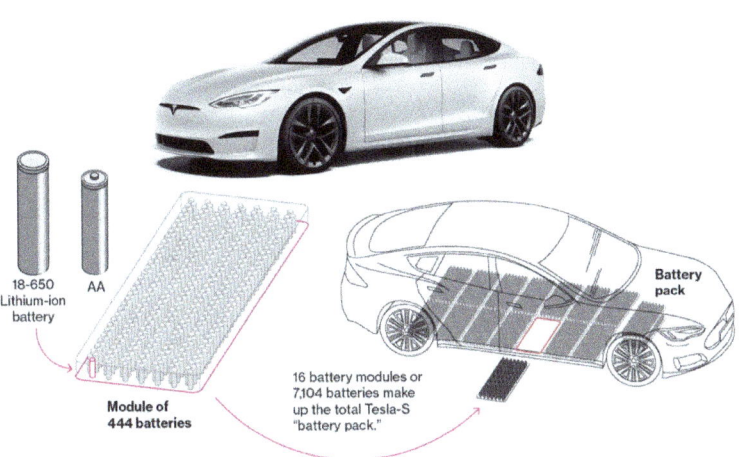

Figure 5. The core technology of Tesla cars is battery technology. The development of battery materials that can charge faster and store greater amounts of electricity is essential.[10]

parents' generations, design and assembly, or integration of small parts, were important. However, development through these processes is now reaching its limits, and the creation of new materials can solve this problem. Even though it is an electric vehicle, its design is not much different from that of a conventional oil-powered vehicle. The appearance of most cars, including their four wheels, two rear-view mirrors, and interior, is not much different. The biggest difference between oil vehicles and electric vehicles is the presence or absence of large capacity batteries, and the key to making these batteries is the development of materials that can charge electricity quickly and in high quantities (**Figure 5**).

To provide another example, the size of transistors in computer central processing units (CPUs) has been reduced through years of

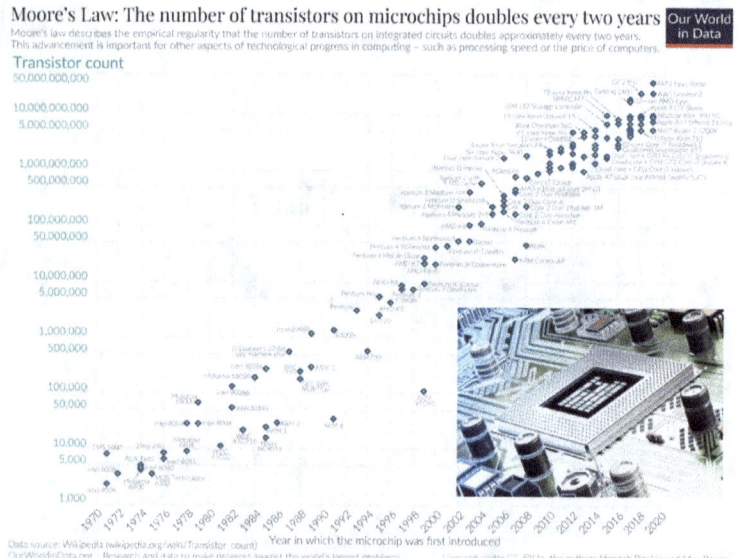

Figure 6. Current computer central processing units have nearly 50 billion transistors integrated. Each transistor is about 10 nanometers in size (nanometer (nm) = 1 billionth of a meter), which is about 10,000 times smaller than the thickness of our hair. Making transistors smaller than this is very difficult with current technology, so the development of better computer chips is more difficult than before[11,12]

integration technology, but has nearly reached the limit of what humans can construct (**Figure 6**). When Seok-Woo was in the middle school (1990s), computer speeds increased at a tremendous rate every year. The computer that people bought a year ago was likely to have insufficient specifications to play new games. As part of a generation that has experienced first-hand the amazing speed of computer development, Seok-Woo feels that the speed of current computers is not improving much in these days. The transistor needs to be made smaller and directly integrated, but the size of the transistor is already so small that it is not easy to reduce it much more. Therefore, a lot of investment is being made in the development of different type of computers, such as quantum computers (**Figure 7**). To develop these next-generation computers, the development of new materials that change the existing paradigm is essential.

We believe that the development of civilization in our generation compared to previous generations, and in future generations compared to our generation, depends more and more on what materials can be developed. We wrote this book because we believe that we are all living in an era where we need to have common sense about what materials science is and how materials scientists create materials. We hope that by reading this book, a reader (you!) will become familiar with the increasingly important field of materials science and engineering.

Google Sycamore Quantum CPU

10 mm

Figure 7. Google's Sycamore quantum computer. Quantum computers that operate on the principles of quantum mechanics, the laws of physics in the microscopic world, can perform certain calculations much faster than the supercomputers currently available to humans. In 2019, a quantum computer using superconductors developed by Google surprised the world by showing that it could perform mathematical calculations in just 200 seconds that would take existing supercomputers 10,000 years. The downside of quantum computers is that quantum circuits can only operate at extremely low temperatures (almost absolute zero, or -273 degrees Celsius). The development of new materials is essential for easy use at higher temperatures.[13]

Good products can only be developed using good materials. Therefore, the role of materials scientists is becoming increasingly important. That's probably why Elon Musk recommended undergraduate students to take the Materials Science 101 class!

Elon Musk ✓ ✗
@elonmusk

Take Materials Science 101. You won't regret it.

7:52 PM · Sep 9, 2022

9,255 Reposts   **1,478** Quotes   **133.8K** Likes   **3,350** Bookmarks

💬   ↻   ♡   🔖 3.3K

(blank page)

Ch. 1. What is materials science?

What is materials science? Materials science is the discipline that studies how materials with 'useful properties' are made and explores the principles by which those properties appear. Material properties are values that numerically express the phenomenon or behavior that a material exhibits when a certain stimulus is applied. When people are pinched (stimulus), they scream in pain (behavior). A weak child may scream and cry, but a very strong UFC athlete may remain silent as if nothing happened. Even though the same stimulus is given, each person reacts differently, and the degree of that reaction can be quantified in decibels of screaming.

When materials are stimulated in a certain way, they show different degrees of behavior depending on the type of material. The degree of this behavior can be quantified. For example, when voltage (stimulus) is applied to a material, electricity can easily pass through a material like copper (behavior), but struggles to pass through a material like glass. Thus, copper has a higher electrical conductivity than glass. The value of electrical conductivity is a measurable quantity that indicates the degree of electrical conduction. Basically, materials scientists create materials, measure their properties, and study the fundamental reasons for why the materials exhibit those properties. The research results are used to further improve the material properties, ultimately leading to an improved material (**Figure 8**). For example, if a material is very hard and does not break easily, a materials scientist uncovers the scientific principle behind this and uses that principle to create a stronger material. We believe that, after continued research like this, we may someday produce materials as strong as Iron Man's armor or Superman's clothes!

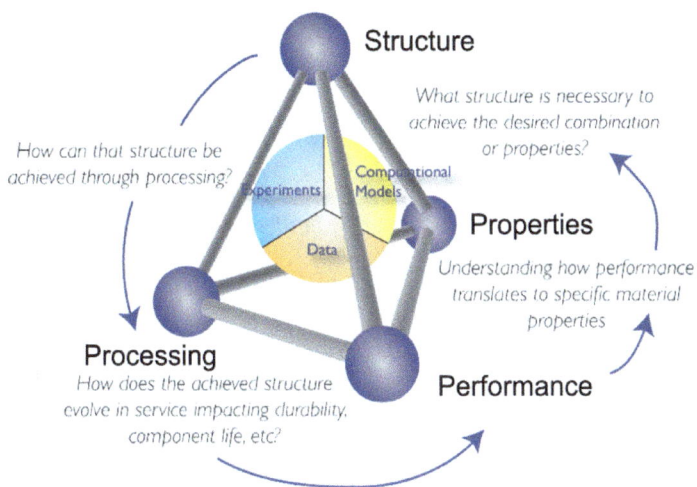

Figure 8. Materials Science Tetrahedron. This is a diagram that any material scientist must memorize. Once materials scientists process materials, they analyze the structure and measure the properties. At the same time, the performance that the material can show when used in applications is also evaluated. By understanding the correlation between the structure and properties of materials, methods to improve the material properties are discovered. Then, the materials are remade using those methods, and structural analysis, material property measurement, and performance evaluation are performed again. That is, a materials scientist develops materials by evaluating the elements of the Materials Science Tetrahedron (structure-property-processing-performance relation) in various ways.[14]

What's important is that materials scientists do not create the Iron Man's armor itself, they create the 'materials' needed to develop the armor. It is very important to understand this distinction, as materials scientists use physics and chemistry to create necessary materials for use by other engineers. These engineers create robots, airplanes, computers, cars, buildings, bridges, and other products. They need better and better materials to make better and better products, and the people who create these materials are the materials scientists. This means that materials scientists must have knowledge of the natural sciences (physics and chemistry), and at the same time, must be able to think about how to make our lives better with the mindset of an engineer.

Therefore, materials science serves as a bridge between science and engineering. This was why we both chose the materials science department when we graduated from high school. We thought it was an interesting discipline that allowed us to work with science, which stimulated our endless curiosity, and engineering, which allowed us to create things that contribute to the development of the world. Many universities have a department of Materials 'Science' and 'Engineering'. Unlike other engineering fields, it incorporates both science and engineering in its name. Thus, it is a field that combines the fun of science, and the usefulness of engineering (**Figure 9**). We are currently living happily studying materials science, and we dream of middle/high school students having more opportunities to be exposed to our field. Materials science is a very attractive discipline that allows us to explore the mysteries of Nature like a scientist and use that knowledge to develop the world.

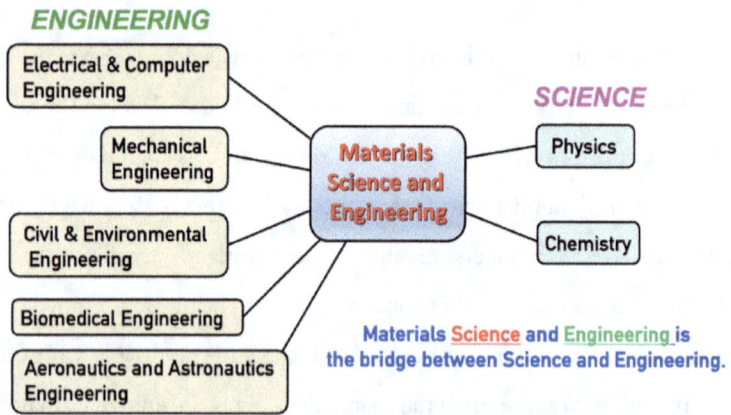

Figure 9. Materials science serves as a bridge between science and engineering. Therefore, it is a very attractive field that can pursue both fun (Science) and usefulness (Engineering).

Then, how do material scientists develop and create new materials? Should we mix random things together endlessly and repeat until good properties are obtained? This is a tedious method that takes too much time. It is much more efficient and less expensive to use scientific theory to design and implement the atomic arrangement of materials. A university's Materials Science and Engineering Department teaches the theories and experimental methods essential for the development of materials. Students create materials directly based on the theories they have learned and study the relation between the internal structure (arrangement of atoms) of materials and their properties. An understanding of this relation allows you to control the properties of materials and learn the processing methods commonly used to obtain improved properties.

Materials science students study the following topics in-depth for about three years (2nd to 4th years) at their university.

- (**Internal structure of materials**) What atomic arrangement and structure does a material have?

- (**Analysis of material structures**) How can we identify the atomic arrangement of a material experimentally?

- (**Materials thermodynamics**) What elements can a material be made of? Are there some materials that are impossible to make?

- (**Materials kinetics**) If you can make the material, how quickly can you make it?

- (**Material properties**) What properties does the material exhibit? – mechanical, electronic, optical, thermal, and magnetic properties of materials

- (**Classes of materials**) How can we classify the materials made?

- (**Material innovation**) Which materials are currently being studied the most?

Through the next chapters, this book will explain these topics as briefly and easily as possible. We hope that reading this book will help you better understand what materials scientists do and inspire you to contribute to the numerous discoveries yet to be uncovered.

Material scientists are people who study the relation between the internal structure (atomic arrangement) of materials and their properties to create useful materials.

Materials science is a discipline that combines science and engineering. So, the more you study it, the more you understand the mysteries of nature (science) and can greatly contribute to the development of our society (engineering).

(blank page)

Ch. 2. The origin of materials on earth

While we were learning materials science, we never learned about where the materials used on Earth originated. This topic is mainly explored by astronomers. We included it as the second chapter because we figured it's important to have a little common sense about the origin of materials before delving into the specifics of materials science. How were all the materials we use formed? How did the elements that form the basis of materials come into existence? Through this chapter, we will answer those questions.

In the beginning, the Big Bang formed the basic particles that make up matter (quarks, electrons) and light. When the universe was very small, it maintained a state of plasma in which elementary particles and light were constantly being created and destroyed. As the universe expanded and its temperature gradually cooled, quarks were stably attached to each other to create protons. Hydrogen was formed when protons and electrons attached to one another via the electromagnetic force. This led the universe to fill with hydrogen, creating the first element on the periodic table.

Hydrogen generated in this way sticks together due to gravity. As the weight of the hydrogen cloud becomes heavier, the gravity and density increase. When the density rises above a certain level, the gravity strongly compresses the hydrogen atoms, causing nuclear fusion, creating helium atoms. During this process, some mass is lost. This lost mass is converted into energy (due to Einstein's $E = mc^2$, where E is the energy, m is the mass, and c is the speed of light) and produces light and heat, resulting in the birth of stars. There are now two elements (hydrogen and helium) on the periodic table (**Figure 10**).

Figure 10. Stars are being created as hydrogen gases come together due to gravity (Carina Nebula taken by the James Webb Space Telescope, NASA).[15]

The energy generated by nuclear fusion raises the star's temperature and causes it to expand. At the same time, gravity causes the star to contract. The balance of these two forces allows the star to maintain a constant spherical shape. At the center of the star, nuclear fusion continues to occur due to strong gravity, producing increasingly heavy elements up to iron, the 26th heaviest element on the periodic table. According to physics, iron is the most stable element. Thus, elements heavier than iron are not created in stars. Therefore, while a star is alive, everything from hydrogen to iron is produced.

As nuclear fusion continues to occur inside a star, heavy elements—which do not easily undergo nuclear fusion—are created. If nuclear fusion does not occur properly and the energy released decreases, so does the star's ability to expand. At some point, when the expansion force becomes smaller than gravity, the star suddenly contracts. This causes a shock wave to be applied to the interior of the star, resulting in

Figure 11. When stars reach the end of their lives, they explode into supernovas and scatter various elements throughout the universe. These elements come together to form new stars and planets. The photo above is the Crab Nebula, showing the remnants left after a supernova explosion (taken by James Webb Space Telescope, NASA).[16]

an explosion that fuses elements heavier than iron. This phenomenon is called a supernova explosion, through which many elements, including iron and the newer and heavier elements, are scattered into space (**Figure 11**). Depending on the weight of the exploded star, a new type of star will form after a supernova explosion; a light star becomes a white dwarf while a heavy star becomes a neutron star or a black hole.

As the scattered elemental gases coalesced under gravity, stars and planets were created, resulting in the formation of the solar system. In the case of a planet, heavy elements gather at the center to form the core, while light elements form the continents and atmosphere. Through various weathering processes over a very long period of time, materials mixed, new materials were formed, and the current shape of the Earth emerged (**Figure 12**). To summarize, the formation and death of stars is essential for the creation of planets, and all the elements and materials used today originated from stars that exploded a long time ago.

Figure 12. Everything on Earth is actually made up of a collection of elements formed by supernova explosions. Therefore, we all originated from a star that exploded a very long time ago.

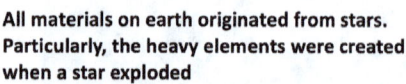

All materials on earth originated from stars. Particularly, the heavy elements were created when a star exploded

So, all of us come from stars!

Ch. 3. Internal structure of materials

Atoms as the basic building block of materials

Those of you reading this book probably have experience playing with LEGOs or similar toys. If you keep attaching and stacking similar-looking blocks, you can create the Millennium Falcon from Star Wars, Hogwarts from Harry Potter, and Elsa's beautiful ice palace from Frozen. In the game Minecraft, you can build your own world by stacking various blocks to build buildings. With LEGOs and Minecraft, various shapes can be created depending on how the blocks, which are the basic structural units, are combined. In fact, making materials follows a similar principle (**Figure 13**). The blocks that materials scientists use aren't blocks, they're atoms of the elements that exist on earth. In short, the development of new materials involves combining atoms—the smallest unit of various

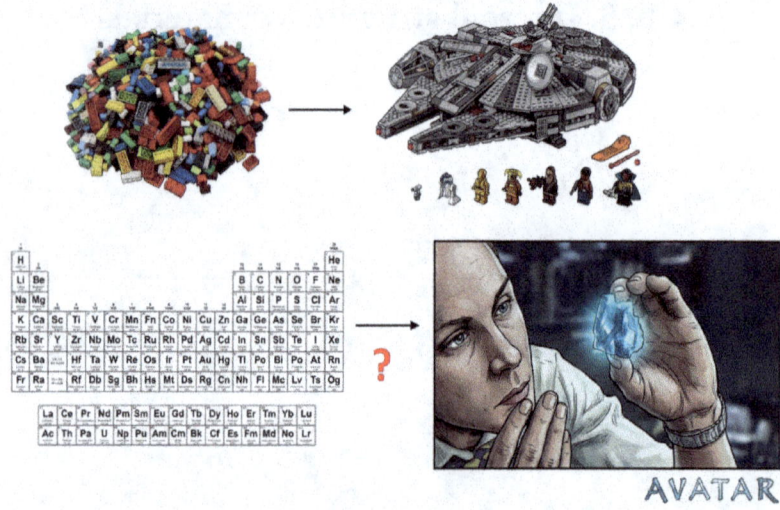

AVATAR

Figure 13. Just like combining blocks to make something in Lego or Minecraft, materials scientists create materials by combining atoms, the basic units of each element in the periodic table. Of course, atoms are too small to accurately pick up and combine them one by one, but their position can be adjusted through various methods so that the atoms as a whole have a certain arrangement. By creating arrangements of atoms that did not exist before, new materials can be developed. (**Avatar, 20th Century Fox**)[17]

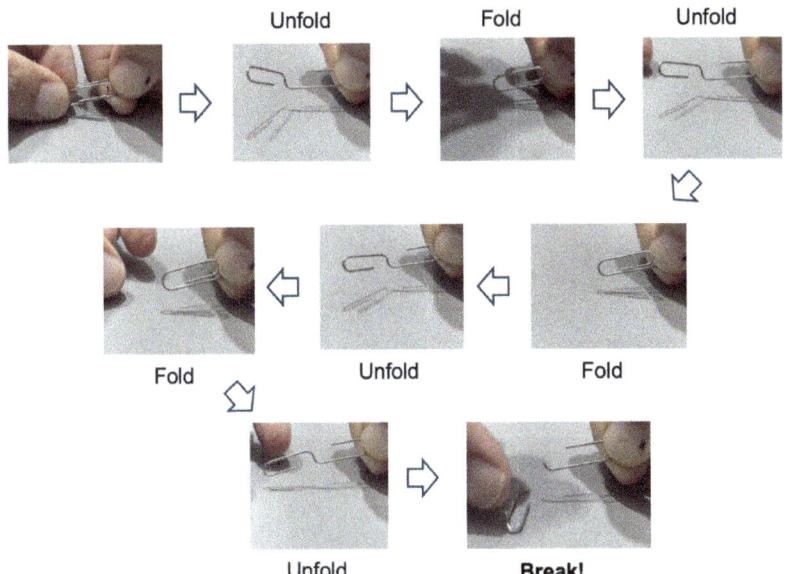

Figure 14. If you bend the steel clip several times, it will eventually break. This is because the atomic arrangement in the bent part gradually changes from a soft/deformable atomic arrangement to a hard/brittle atomic arrangement through the repeated deformation. Because this is an atomic-level phenomenon, it cannot be seen with the naked eye, and the different atomic arrangement can only be seen using a special microscope (Transmission Electron Microscope, See **Chapter 4**).

elements in the periodic table—in various ways. Because the properties and performance of a new material depend on the specific atoms the material is made of and the arrangement of those atoms, materials scientists must be able to understand and analyze the **composition of the material** and **the arrangement of the atoms**.

In general, materials are made by mixing different ratios of elements. The easiest way to do this is to melt all the raw materials together, mix them in a liquid state, and then solidify them. At this time, the properties of the materials can vary greatly depending on the mixing ratio, so they must be mixed well in the correct ratio based on scientific theories and experience. However, what's even more fascinating about materials science is that even if the ratio of elements is the same, completely different material properties can be obtained by controlling the atomic arrangement. In the case of a paperclip made of steel, it can return to its original shape if you bend it once or twice, but if you repeat this a few more times, it suddenly breaks (**Figure 14**). At first, when force is applied, the atoms near the folded part may move around well allowing for a change in the shape of the paperclip. However, as bending is repeated, the atoms near the bend shift to an atomic arrangement that makes movement difficult. Eventually, when the atoms cannot move at all, bending becomes impossible, causing the chemical bonds between the atoms to break and the paperclip to snap. In other words, repeated bending causes the folded part of the paperclip—which starts out with a soft and easily deformable atomic arrangement—to change into a material that has a hard and brittle atomic arrangement (like glass). During the bending process, the material composition didn't change, as we never mixed or subtracted any elements. To put it simply, the mechanical behavior of the paperclip was changed by changing the arrangement of the atoms without altering the ratio of the elements that make up the material. Controlling the properties of materials by adjusting the atomic arrangement is a core topic in materials science. It took Seok-Woo more than 10 years to realize the importance of this idea. We hope that reading this book allows you to understand this much quicker than he was able to!

Basic arrangement of atoms

Materials are broadly divided into crystalline materials and amorphous materials (**Figure 15**). Crystalline materials refer to materials

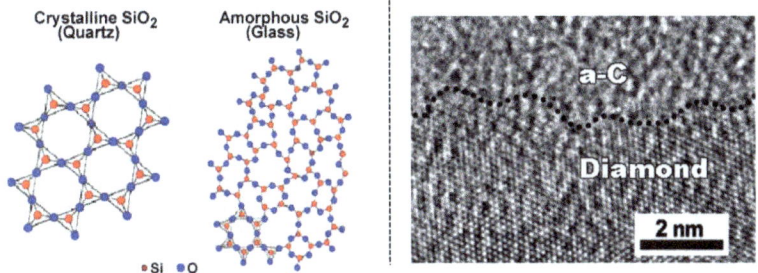

Figure 15. Crystalline and amorphous materials. Crystalline materials have atoms arranged periodically, but amorphous materials have atoms arranged randomly. (Left) When silicon and oxygen are arranged periodically, it becomes quartz, but when they are arranged randomly, it becomes glass. Quartz and glass are equally optically transparent, but their electrical and mechanical properties are different. (Right) Amorphous carbon and diamond (crystalline carbon) are stuck together. This photo was taken at high magnification using a transmission electron microscope, allowing an atomic-level photo to be obtained. The white dots visible on the diamond side correspond to carbon atoms, and you can see that the carbon atoms are arranged periodically. However, the structure of the amorphous carbon portion (a-C) is not clearly visible because the atoms are randomly arranged like a liquid. [18,19]

in which atoms are lined up neatly and repeatedly, while amorphous materials refer to materials in which atoms are randomly arranged. Most metals and ceramics are crystalline materials, while polymers (plastics) and glasses are amorphous materials. In the case of amorphous materials, it is very difficult to understand microstructure due to the random arrangement of atoms. In the case of crystalline materials, the structure is theoretically well established due to the beautiful periodic arrangement of atoms. Thus, this chapter deals primarily with the atomic arrangement of crystalline materials.

According to crystallography, there are 14 basic **lattice structures** in three dimensions. A large lattice structure can be created by continuously connecting these 14 basic lattices in a three-dimensional

31

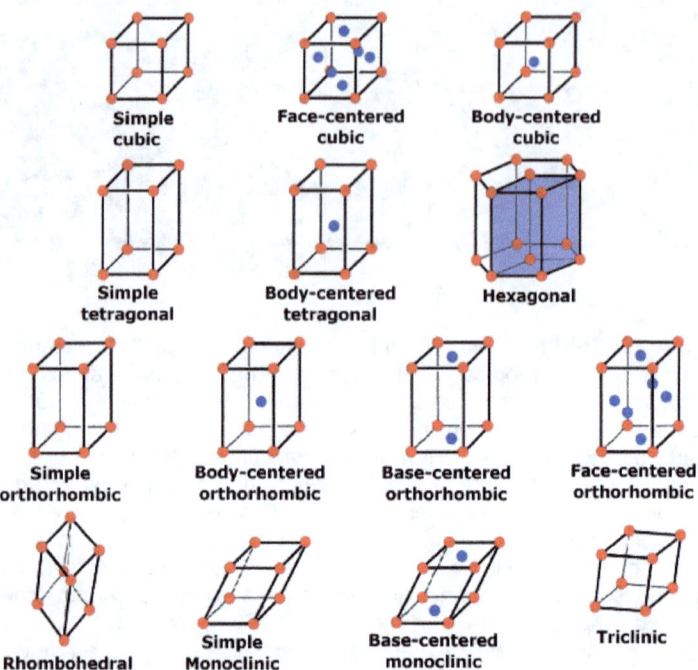

Figure 16. 14 mathematically proven possible atomic arrangement structures. By repeatedly attaching this basic lattice structure in three dimensions and adding atoms into each lattice point, possible atomic arrangements can be created. All crystalline materials have the above 14 basic lattice structures. These basic lattices contain lattice points (red and blue dots), which can contain a single atom or a group of atoms.[20]

space (**Figure 16**). This has been mathematically proven—there are no possible basic lattice structures other than these 14. Now, we can implement actual materials by placing atoms in this lattice structure. The simplest way is to place one atom at each lattice point. For example, if an iron atom is placed at each lattice point of a body-centered cubic structure, it becomes the iron that we know. If an aluminum atom is placed at each lattice point of a face-centered cubic structure, it becomes the aluminum that we know. It isn't necessarily the case that just one atom is located at each lattice point. In many crystalline materials, groups of atoms are placed at each lattice point, and the groups of atoms are repeated across the lattice structure. Amazingly enough, a material with a unique atomic arrangement that does not follow these 14 basic lattice structures was discovered. This rare and unique material, called a quasicrystal, exists only in a very small number of materials. The amazing discovery was awarded the 2011 Nobel Prize in Chemistry (**Figure 17**).

Iron has a body-centered cubic structure at room temperature but has a face-centered cubic structure above 907° C. We were amazed when we learned this as undergraduate students (**Figure 18**). The fundamental change in the state of matter is intuitive. When iron gets too hot, it changes to a liquid or gas. According to physics, the vibration of atoms becomes stronger as the temperature rises. When the atoms tremble too much, their chemical bonds break, and they flow (liquid) or fly apart (gas). However, in the case of iron, the moment the temperature exceeds 907 °C, hundreds

Figure 17. (Left) Photo of Ho-Mg-Zn quasicrystal, (Right) Atomic model of an aluminum-palladium-manganese quasicrystal surface. In the case of crystals, atoms or groups of atoms repeat in a certain direction, and the same structure can be observed if the structure is rotated about a certain axis. However, in the case of quasicrystals, the same structure can be repeated with respect to rotation, but the atomic arrangement does not repeat in a direction. Dr. Dan Sechtman received the Nobel Prize in Chemistry in 2011 for first discovering the existence of quasicrystals.[21,22]

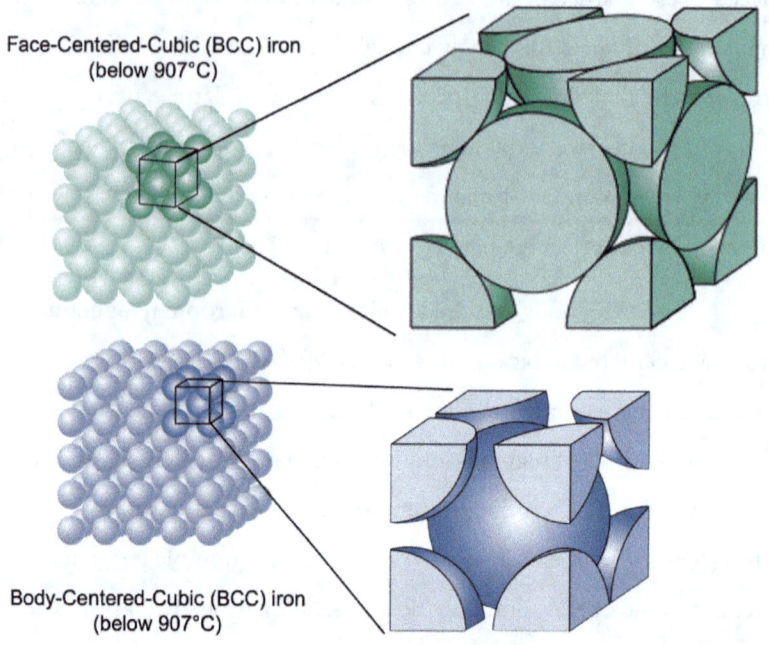

Face-Centered-Cubic (BCC) iron
(below 907°C)

Body-Centered-Cubic (BCC) iron
(below 907°C)

Figure 18. Two different structures of iron. Below 907 degrees Celsius, it has a body-centered cubic atomic arrangement, and above 907 degrees Celsius, it has a face-centered cubic structure.[23]

of billions or trillions of atoms suddenly change their atomic arrangement from a body-centered cubic structure to a face-centered cubic structure while preserving their solid state (**Figure 18**). This is truly amazing, isn't it? Since the face-centered cubic structure is denser than the body-centered cubic structure, the iron atoms are much more closely packed together. If there is a long wire, its length becomes instantly shorter when the temperature exceeds 907°C. In addition to the sudden change in dimension, when the periodic arrangement of atoms changes, the properties of the material also change. Thus, it is incredibly important for materials scientists to understand the arrangement of the atoms in order to understand the behavior of developed materials.

Modifying the atomic arrangement (Microstructure)

What perspective do materials scientists have about materials development? In general, materials scientists think that changing the arrangement of atoms means creating a different material. In other words, after obtaining an atomic arrangement with one of the 14 lattice arrangements seen in the previous section, we want to control the properties of the material by gradually changing the atomic arrangement. Therefore, materials scientists must be able to understand what is happening at the atomic level. When Seok-Woo explains materials development to students, he often draws a picture of how the atoms are arranged on the blackboard and then show them how differently the atoms can be positioned. Studying materials science requires an understanding of a small-scale microscopic world that we can't see in our day-to-day lives. Seok-Woo found that students sometimes find this

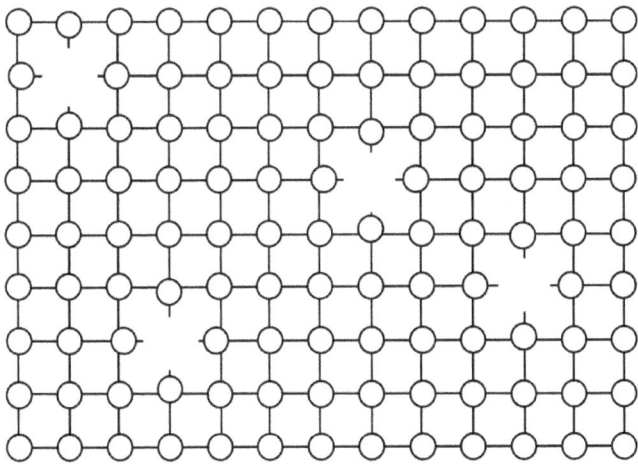

Figure 19. The space where the atoms escape is called a vacancy. The number of vacancies increases as the temperature increases. Vacancy is an essential defect when allowing other atoms of similar size to diffuse into host materials

challenging when they first encounter materials science. However, it is very important to understand the arrangement of atoms one-by-one by drawing pictures. In this section, let us explain this process step-by-step.

First, vacancies can be created by removing some atoms (**Figure 19**). The existence of vacancies is essential for mixing materials in the solid state. When mixing other elements into a silicon wafer to make a semiconductor, there must be empty space. Without a vacancy, it is almost impossible for some important elements to diffuse into the host material. We will discuss the role of vacancies in diffusion in detail later when we discuss material kinetics (See also **Chapter 7**).

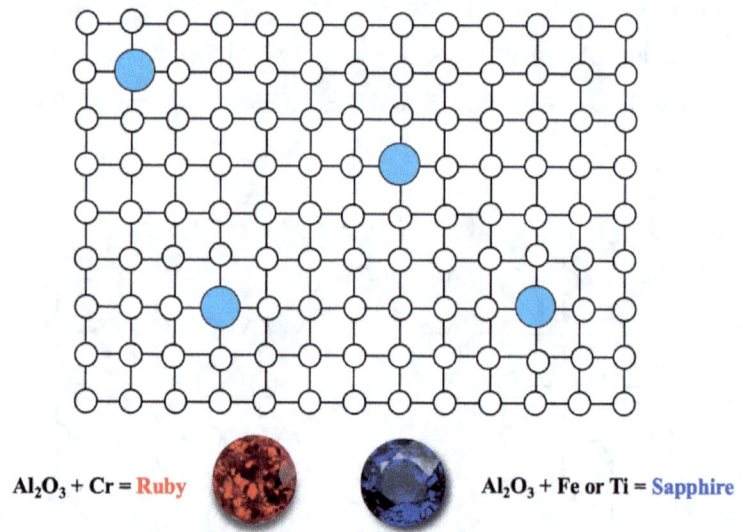

Al_2O_3 + Cr = Ruby Al_2O_3 + Fe or Ti = Sapphire

Figure 20. When another atom takes the place of an existing atom, it is called a substitutional point defect. A state in which different atoms are randomly mixed as shown in the picture is called a solid solution. A representative solid solution is the semiconductor used in computers. Semiconductors can be made by changing electrical properties by replacing part of silicon (Si) with other atoms. As another example, if some of the aluminum in alumina are replaced with chromium, it becomes red ruby, and if it is replaced with iron or titanium, it becomes blue sapphire. So, the optical properties of a material can also be changed by substitutional point defects.

Second, existing elements can be replaced by new elements (**Figure 20**). This significantly changes the properties of the material. For example, alloys, in which some elements are substituted by other elements, are much stronger than high-purity metals. Even the small addition of other elements to high-purity aluminum can result in the material becoming dozens of times stronger (in other words, dozens of times the force is required to break it). This results in the formation of an alloy: a mixture of two or more elements consisting of at least one metal. The mass density of an aluminum alloy is not much different from that of pure aluminum, which is why aluminum alloys are very light and incredibly strong. Because airplanes must be strong and light enough to fly, aluminum alloys are used as structural materials. Some say that without aluminum alloys, the airplanes used today would not exist. In addition to changes in strength, the color of the material may change due to the substitution of other elements. For example, alumina (Al_2O_3) is a colorless and transparent material, but if iron is added to some of the lattice positions occupied by aluminum, it becomes a blue sapphire color. If chromium is added, it becomes red ruby. In other words, the optical properties of materials can be changed by elemental substitution (**Figure 20**). Transistors in semiconductors are made in a similar way. By carefully mixing different impurities (boron or phosphorous) into a silicon wafer, a path can be created, allowing electrons to flow. By allowing or preventing current from flowing, logic calculations can be performed by representing this flow as either 1 or 0. This system is known as binary.

Third, very small atoms can be inserted between existing atoms (**Figure 21**). If an atom is very small, it will fit between the atoms around it without having to replace them. For example, small atoms like carbon

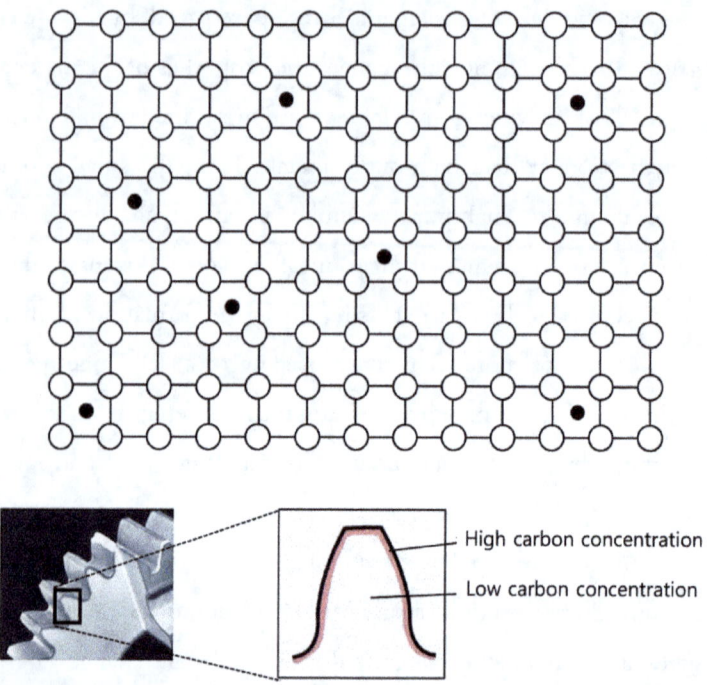

Figure 21. A very small atom, which exist between the lattice of an existing material, is called an interstitial point defect. Gears must have a wear-resistant surface because they undergo surface wear over a long period of time. The surface can be strengthened by infiltrating carbon atoms into it. Because carbon is a very small element, it can penetrate between the lattices of the material that makes up the gear.[23]

can penetrate other materials very easily, inserting themselves in empty spaces. In this way, the surface of iron can be strengthened by adding small amounts of carbon into the surface. In metal gears, wear can be reduced by putting small amounts of carbon into the surface, strengthening the material. All high-strength iron alloys contain carbon.

Fourth, it is possible to create a structure that contains an additional atomic plane (**Figure 22**). This structure is called a dislocation. Dislocations move when a force is applied to the material. The movement

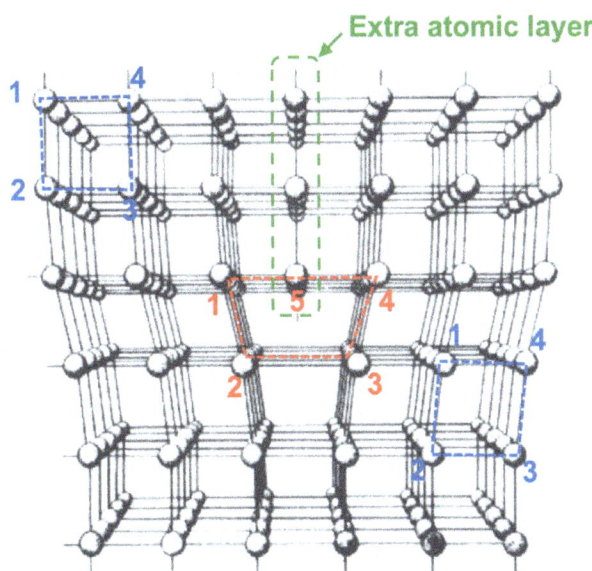

Figure 22. The region indicated by the red broken line has a local lattice of five atoms, while all other parts have a lattice of four atoms. Therefore, the part made up of five atoms is different from other parts, and this special structure is called a dislocation. This structure sometimes looks as if it was made by inserting extra row of atoms (indicated by a green broken line) into the material.

of numerous dislocations allows the material to permanently change its shape (generally referred to as plastic deformation). Additionally, plastic deformation of materials can create new dislocations.

The bending of the paperclip, as discussed earlier, is closely related to the motion of dislocations (**Figure 23**). As the paperclip is bent, the number of dislocations that allow the material to permanently deform increases. This results in a permanent change in the shape of the material. Permanent deformation of materials can also create new dislocations. When dislocations are created and continuously duplicated, they impede each other's movement. This makes it more difficult to deform the material. It's just like a traffic jam—when too many cars are on the road—

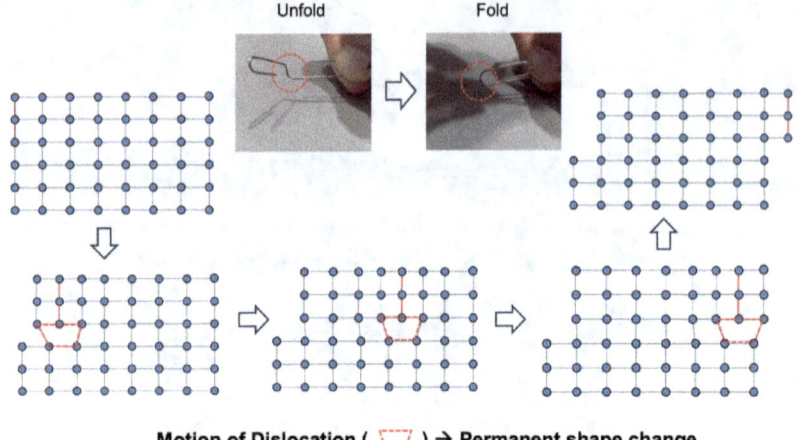

Motion of Dislocation (⬚) → Permanent shape change

Figure 23. When force is applied, dislocations are created and move, deforming the shape of the material permanently. For example, when bending a metal paperclip, if you look it closely (so closely that you can see the atoms through a special microscope), you can see that the shape of the material is changed by the movement of numerous dislocations. To make a strong material, that is, a material that does not easily change shape, it is necessary to hinder the movement of dislocations. In other words, to make a strong metal material like Iron Man's suit, a deep understanding of dislocation motion will be needed.

making it more difficult for any one of the cars to move. This is comparable to an increase in the number of dislocations, which makes it more difficult for any of the dislocations to move. In other words, the clip gradually becomes harder through bending, turning into a brittle material. In movies like The Lord of the Rings, there are often scenes where metal is heated and hammered to make a sword. From a materials science perspective, that hammering is used to add the appropriate number of dislocations into the sword, making it harder.

In fact, Seok-Woo has been researching how dislocation structure affects the strength of materials for the past 20 years. The development of high-strength materials depends on the creation of an atomic arrangement

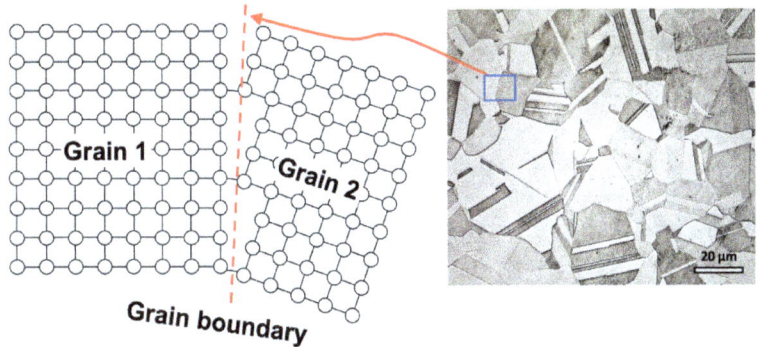

Figure 24. Each grain consists of the same material but bonded into a different orientation. The picture on the right is the grain structure of high purity copper.[24]

that hinders the movement of dislocations. The development of ultra-high strength materials is very important in the military, automobile, and aviation industries. One such example is the development of Iron Man's armor, which ultimately depends on developing and understanding a unique dislocation structure. Dislocations are so important that when Seok-Woo was doing his doctorate at Stanford University (2006~2011), the password for all of the public computers in the Department of Materials Science and Engineering was "dislocation".

Fifth, we consider the structure in which other atomic arrangements connect (**Figure 24**). When crystalline materials change from a liquid to a solid, the phase transformation known as crystallization occurs randomly at various locations in the liquid. As a result, if the material turns completely solid below the melting point, these small single crystals join. Each single crystal is called a grain, and the boundary between them is referred to as a grain boundary. The properties of the material change a lot depending on how many grain boundaries exist. As the number of grain boundaries increases, it becomes more difficult for

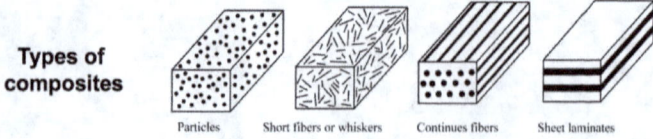

Types of composites

Particles   Short fibers or whiskers   Continues fibers   Sheet laminates

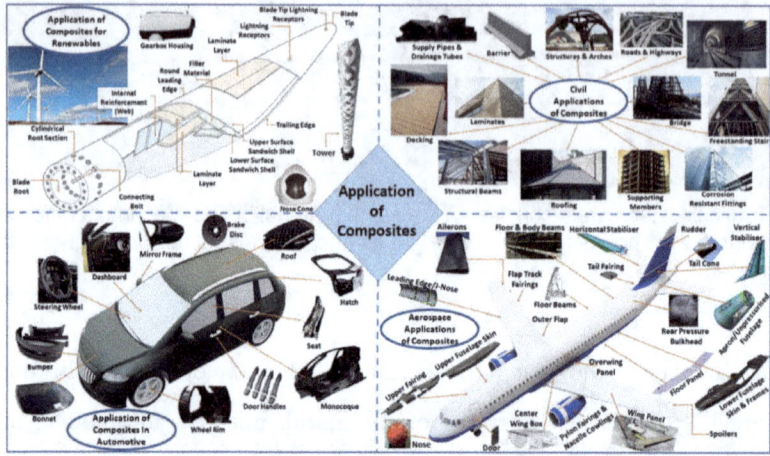

Figure 25. Complex material made by mixing different materials. The advantages of different materials can be combined to obtain optimized properties.[25]

dislocations to move, resulting in an increase in the strength of the material. Additionally, light interacts with grain boundaries, making the material opaque. Grain boundaries also interfere with the flow of electrons and heat, lowering the electrical and thermal conductivities of the material. These material properties will be discussed in further detail in **Chapter 8**. At high temperatures, atoms can flow through grain boundaries. This means that it is easier to change the shape of materials at high temperatures. Thus, the engine blade of a plane must be made of a large "single crystal" that has no grain boundaries to avoid the change that would take place in a hot engine. To make multiple computer chips on a single silicon wafer, the atomic arrangement of the silicon wafer must be uniform. This is why the wafer is made of large single crystals.

Sixth, you can create a composite material that mixes multiple different materials (**Figure 25**). Composite materials are made to take advantage of their constituent materials. For example, if you mix a light plastic and a strong carbon fiber, it becomes a lightweight high-strength composite that can be used in aerospace and military industries.

Amorphous Materials

Not all materials found in our day-to-day lives are crystalline. Some exhibit a random atomic arrangement (**Figure 15**). These materials are known as amorphous materials. In a crystalline material, there is long-range order—an array of atoms is repeated over a long distance range. In an amorphous solid, there is no long-range order—the atomic arrangement is not repeated, and the atoms are randomly ordered. There are many different types of amorphous solids. The glass windows in your house are an amorphous material. The rubber used in your car tires is amorphous. The LEGO bricks that you've played with all your life, also amorphous. Amorphous materials also play a critical role in our lives and are an important topic of study for materials scientists.

The most well-known amorphous material is glass (**Figure 26**). Glass is typically made of silicon dioxide (SiO_2), and is formed by heating sand until it melts, and then rapidly cooling. This rapid cooling prevents the silicon dioxide from forming a crystalline structure. Imagine you and your friends are standing at the front of your classroom. The teacher gives you 20 seconds to get to your chairs. This should be enough time for you to get to your chair and sit down—an ordered arrangement. Now imagine your teacher gives you 3 seconds to get to your chair. Maybe one friend

Figure 26. Stained glass is regularly used to make gorgeous windowpanes for churches and cathedrals. To form glass panes, silica must be melted down. To give it its beautiful color, metallic oxides are added to the molten mixture. The cooled glass is then formed to make beautiful panes, like those seen above.

makes it to his chair, but everyone else is left standing in a random place—there is no order. This is how amorphous glass is formed. You and your friends aren't given enough time to form an ordered structure, just as liquid atoms aren't given enough time to form a crystal structure. This is also why glass is transparent, as the lack of order allows for light to easily pass through.

Another important class of amorphous material is polymers, which will be discussed in greater detail in **Chapter 9**. Polymers are made of long chains of molecules. These molecules can take many different forms such as lines, branches, or networks, but are based entirely around carbon atoms (**Figure 27**). You can think of it like spaghetti. When it is cooked and put in a bowl, it becomes randomly jumbled and tangled,

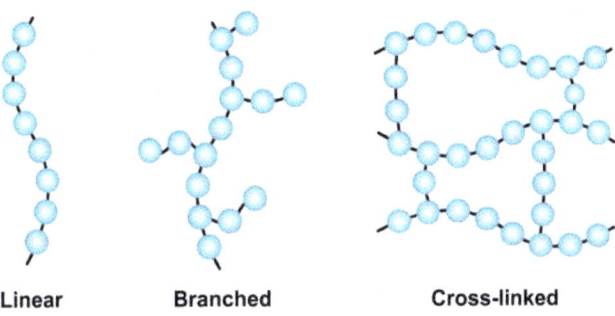

| Linear | Branched | Cross-linked |

Figure 27. The three different polymer structures—linear, branched, and cross-linked—are different due to the structure of the building blocks. Though each of these structures may consist of the same elements, differences in the way the building blocks are arranged results in differences in properties.

resulting in an amorphous structure. This amorphous structure gives polymers properties like flexibility and transparency and is why we can see through and easily manipulate things like plastic wrap.

Like crystalline materials, glass can be modified using additional elements. By adding things like soda and lime, materials scientists can disrupt the network structure. This has vast implications on the properties and formability of glass, making it easier for the material to both melt and be shaped. If you've ever accidentally hit a baseball through a window, you know that it's incredibly fragile. If that window was made of a plastic, however, it would be incredibly flexible. Even though two materials are amorphous, differences in structures (a glass network vs. a polymer chain) result in drastic differences in properties. This further underscores the important relationship between the structure and properties of materials.

(Dr. Seok-Woo Lee's Note):

"When I was young, I was very interested in the birth of the universe and the creation of stars and planets. Therefore, I would have liked to study astronomy just as much as material science. If I didn't choose the Department of Materials Science and Engineering, I probably would have chosen the Department of Astronomy. But I quickly realized that material science and astronomy really aren't that different. After the Big Bang, the stars and planets were formed, and on our planet, various materials emerged by arranging different atoms together. Furthermore, by some chance, protein molecules formed the first life on Earth. Thus, I feel that the evolution of the universe is material science itself! Through various evolutionary processes, the present flora, animals, and humans emerged. Intelligent humans are able to arrange atoms and make new materials (perhaps not present in the universe), making a better world. The most interesting part of materials science is the creation of new materials using the atoms given to us by the stars. I am still amazed to this day by the close relationship between the formation of matter and the evolution of our universe."

Making a material involves either changing the composition of the material or changing the atomic arrangement. In order to understand materials more deeply, therefore, it is very important to imagine the internal structure of materials from an atomic-level perspective.

Even without changing the composition of a material, its properties can be greatly changed by changing the arrangement of its atoms, and this is the most important part of studying materials science.

(blank page)

Ch. 4. Analysis of material structures

Imagine that a meteorite suddenly falls to Earth, and the materials within it have special properties that have never been seen before. Maybe it floats in the air like in the movie Avatar, or constantly releases a strong form of energy. What kind of information do we need if we want to make these mysterious materials by ourselves? Assuming that most of the universe is made up of the elements of the periodic table, it is necessary to understand what elements are mixed and how the atoms are arranged in the material. It's just like building something with LEGOs, if we know the pieces used in an arrangement, we know the final product. As mentioned in the previous chapter, the properties of materials are strongly related to the composition of elements and the atomic arrangement. We can uncover the proportion of elements and the arrangement of atoms using various types of analytical equipment. This allows us to not only understand the properties, but also to produce the same materials.

X-ray diffraction

The atomic arrangement (crystal structure) of a material can be obtained through an indirect method that doesn't require us to see the arrangement of the atoms directly. For example, if you shine a light or electron beam at a material, you can find information about the atomic arrangement using the light or electron beam that bounces back.

The method most commonly used by material scientists to uncover the atomic arrangement of materials is X-ray diffraction (**Figure 28**). If we shoot an X-ray onto a material at various angles, diffract at some special 'incident' angles. The intensity of the diffracted 'incident' X-rays can be drawn on a graph. Using this graph, special patterns can be

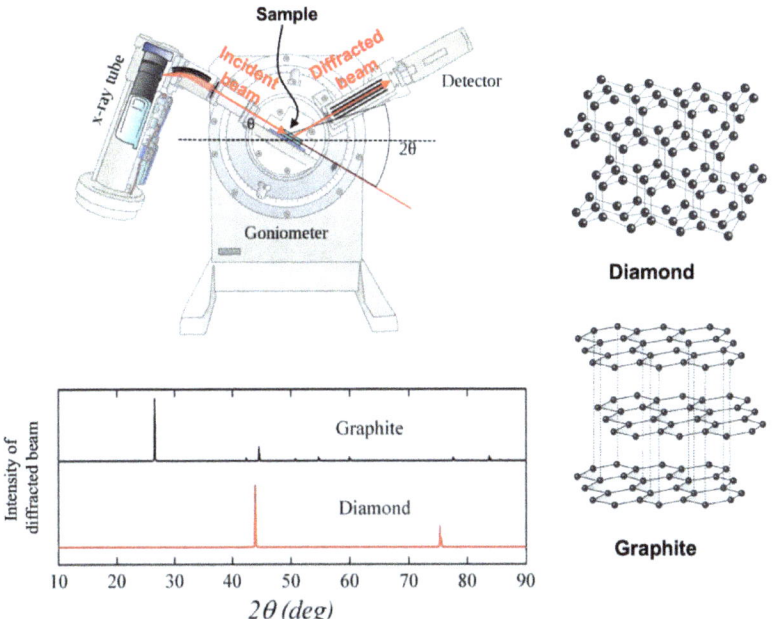

Figure 28. (left top) Schematic diagram of an X-ray diffraction instrument. The X-rays diffracted from the material enter the detector. The intensity of the diffracted X-ray beams is measured at multiple incident angles (θ). According to the structure of the material, the X-rays are diffracted only in a special incident angle. (Right) The graphite and diamonds are made of the same carbon atoms, but their arrangement shows different diffraction patterns (left bottom).[26,27]

observed. Since this pattern changes according to the atomic arrangement, you can uncover atomic arrangements based on known patterns. X-ray diffraction is theoretically well established, so the diffraction pattern coming out of any given crystal structure can be predicted using calculations. This calculated result can be compared with the diffraction pattern measured by the experiment, allowing materials scientists to uncover the atomic arrangements present in a material.

Optical microscopy (magnification: 1 ~ 1,000 X)

If you look at the clothes that we wear from afar, the fine knitted seams are not visible to the human eye. The eye does, however, observe that each area is attached to patches of different colors. Similarly, if a material is observed at a low magnification, the arrangement of the atoms is not visible, but the patches that consist of different atomic arrangements can be observed. This is referred to as the microstructure of the material.

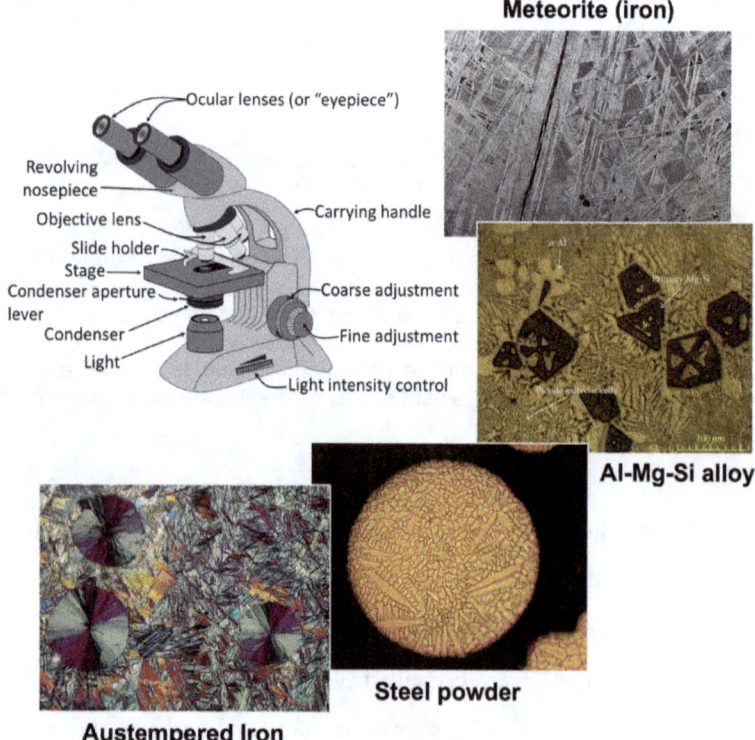

Meteorite (iron)

Al-Mg-Si alloy

Steel powder

Austempered Iron

Figure 29. The length scale of microstructures is too small to see with the naked eyes. Thus, the microstructure of the material is usually visible only under the optical microscope. After polishing, some chemicals can be used to add colors onto the surface, allowing you to take beautiful pictures. Because the microstructure of materials is not something that we are familiar with, I often get a mysterious feeling whenever I acquire a new microstructure image.[28-32]

Even if the arrangement of the atoms cannot be seen in detail, it is important to observe a materials microstructure because the properties of a material are closely related to its microstructure.

The first step in observing the microstructure of a material is the use of an optical microscope. When Seok-Woo was young, not many people had an optical microscope at home, so he could only use it during science class. Today, we can easily find a good quality optical microscope that can be used with USB on your computer at a low price. Optical microscopes are used when the magnification required is in the range of 10 to 1000 times. The visible light that we see with our eyes is expanded through the glass lens, allowing us to observe the microstructure seen on the surface of the material (**Figure 29**).

Usually, the surface of a material is microscopically bumpy and dirty. Thus, light is randomly scattered, and the microstructure is hard to

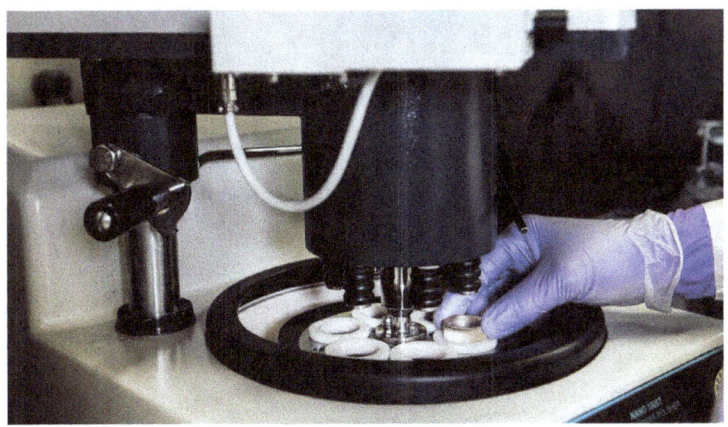

Figure 30. Polishing work. (Step 1) Cut the materials into the right size first, (Step 2) mount a sample into a holder for easy gripping (Step 3), and grind until the surface looks like a mirror with fine sandpaper or alumina micropowders. (Step 4) After that, if we corrode the surface slightly with chemicals, the microstructure will be visible. In order to obtain a beautiful microstructure picture, you need to know the right corrosion conditions.[33]

see. Therefore, materials scientists must polish the surface before performing optical microscopy. First, the materials must be cut and the surface made flat and clean by grinding with fine sandpaper. Then, the surface is modified using chemicals, and the microstructure becomes visible. This series of steps is called 'polishing' (**Figure 30**). When Seok-Woo was an undergraduate student, there were some professors who judged students' overall experimental ability by the quality of their microstructure images. The student who brought the most beautiful microstructure image received an A. This means that polishing is one of many experimental skills that a materials scientist must learn.

Scanning electron microscopy (mag.: 10~ 100,000 X)

The scanning electron microscope (sometimes abbreviated "SEM") obtains an enlarged image compared to that of an optical microscope. As the name suggests, this is done without the use of visible light, instead using an electron beam (**Figure 31**, **Figure 32**). When accelerated electrons hit an object, an electron in the object pops out. These secondary electrons allow us to get an image of very high magnification (up to one hundred thousand times) of the surface topography—hills, bumps, and texture. The scanning electron microscope still cannot see the arrangement of atoms directly but has the advantage of being able to observe microstructures in a fine area. If you need to see very small things that can't be seen with an optical microscope, a scanning electron microscope should be used.

Another good thing about scanning electron microscopy is that the element ratio can be obtained. When accelerated electrons bombard

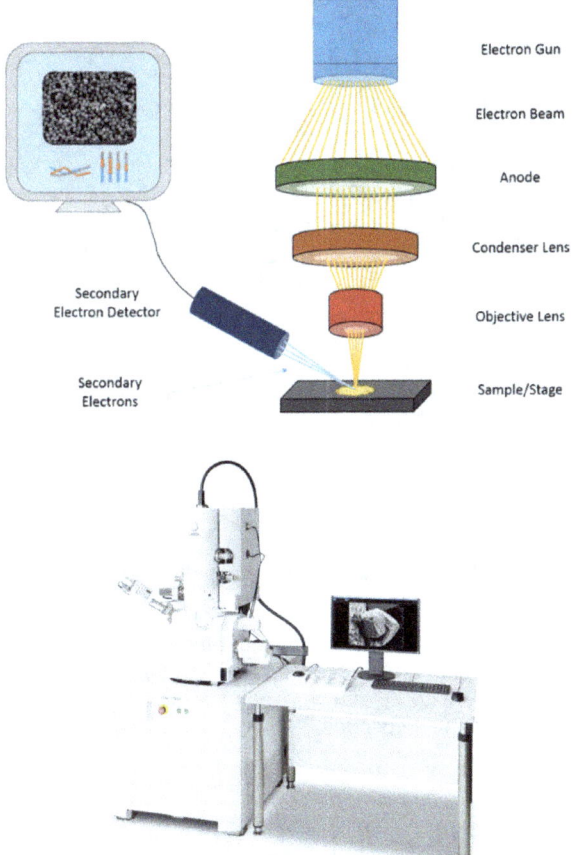

Figure 31. Scanning electron microscopic schematic and photo. If you focus on the accelerated electronic beam (yellow line) with a lens using the magnetic field, the secondary electrons on the sample surface will pop out.[34,35]

the atoms in a material, the impacted atoms release X-rays with different wavelengths. If we measure the wavelength of these X-rays, we can see what elements are present in the material. If several elements are mixed, the scanning electron microscope can be used to determine the distribution of the elements within the microstructure (**Figure 33**).

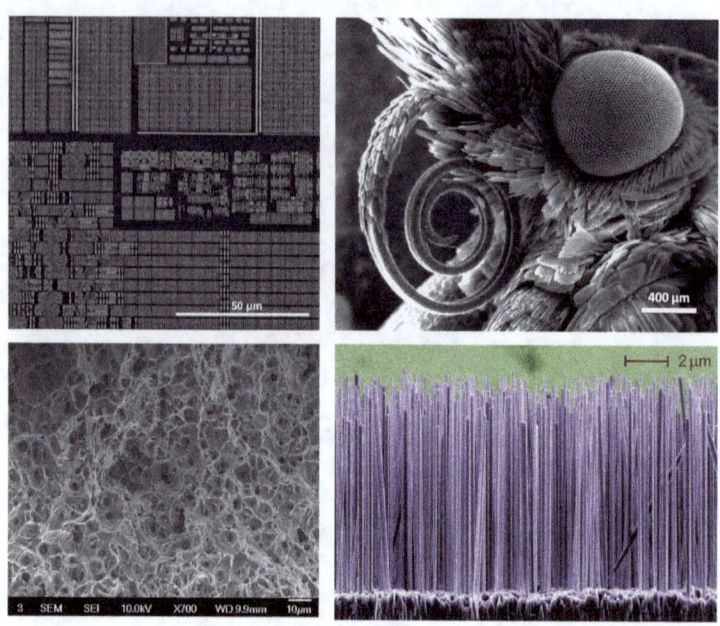

Figure 32. Photos taken with scanning electron microscope. Computer's CPU (upper left), moth (upper right), broken surface of aluminum alloys (bottom left), GaN nanowires (bottom right). You can see more details of small parts that are not seen with an optical microscope.[36-38]

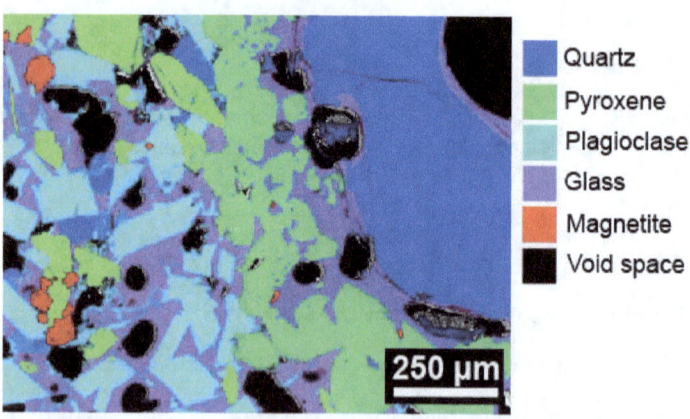

Quartz
Pyroxene
Plagioclase
Glass
Magnetite
Void space

250 μm

Figure 33. Analysis of the components of volcanic rock taken by the scanning electron microscope[39]

Figure 34. Beautiful scanning electron microscope images exhibited at an international conference (Source: Materials Research Society)[40]

In our experience, the scanning electron microscope is incredibly useful for materials scientists for a wide variety of reasons. Because the magnification is so high, people can see things that they would not usually be able to see. At international conferences, material scientists compete by showcasing their best microscopic photographs—it's one of our favorite parts of these events (**Figure 34**).

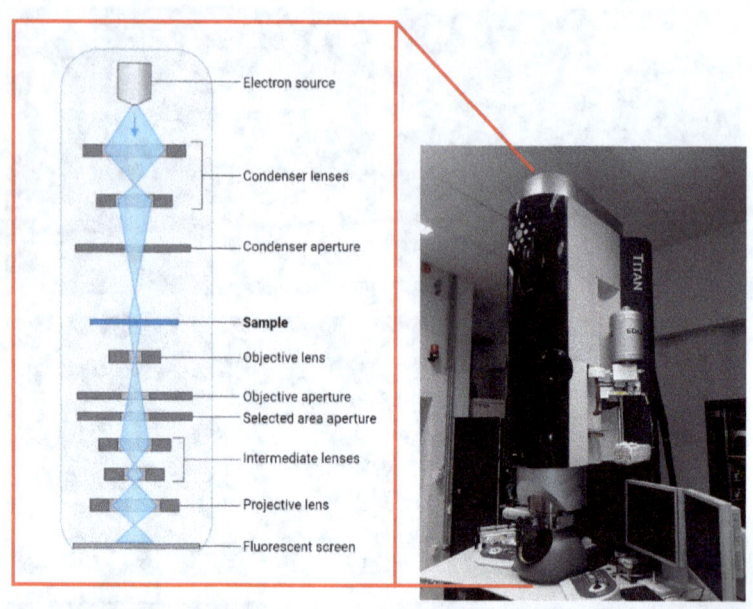

Figure 35. Transmission electron microscope schematic and photo. Unlike scanning electron microscopy, a strong accelerated electron beam passes through the specimen, and this transmitted electron beam delivers the microstructure of the specimen or the atomic array information to the image sensor.[41]

Transmission electron microscopy (up to 10,000,000 X)

The transmission electron microscope is one of the most expensive pieces of equipment used by material scientists (**Figure 35**). Transmission electron microscopy is used to obtain ultra-high magnification images using an accelerated electron beam. If we make the specimen very thin and shoot the strong electron beam toward it, the electrons are able to pass through the material. When the electron beam passes through the material, the electrons interact with the atoms, providing information about the materials microstructure or the arrangement of atoms (**Figure 36**). Optical microscopy and scanning electron microscopy allow researchers to see the surface of the material, while transmission electron microscopy allows researchers to see the insides and obtain images of the hidden microstructure. If the magnification is increased enough, it is possible to see the atomic arrangement directly. The first time looking at the atomic arrangement of a material is unforgettable. It is an experience on another level of nature, and it is a privilege that only material scientists can feel.

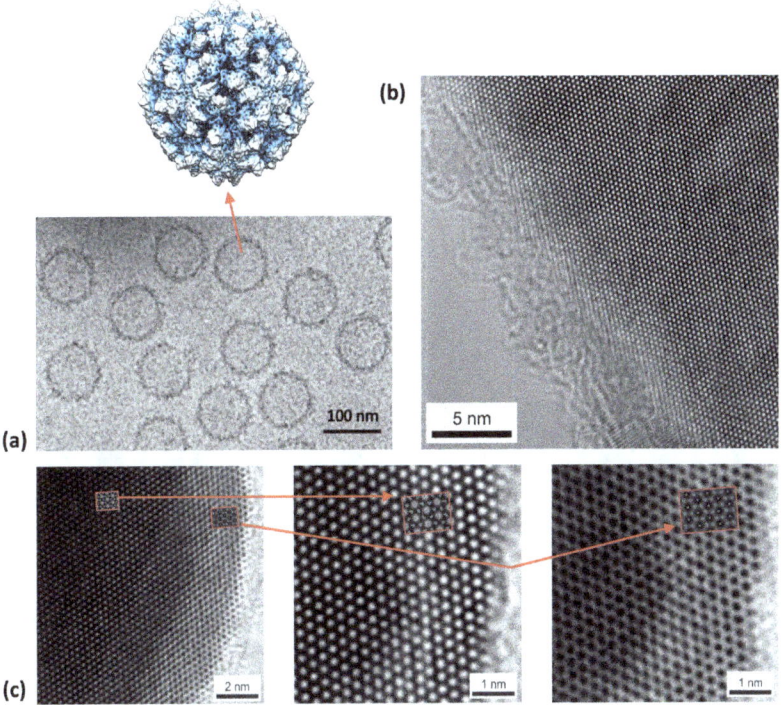

Figure 36. High-resolution (atomic-scale) transmission electron microscope image of (a) Rosellinia necatrix quadrivirus, (b) Pure magnesium, (c) Iron Oxide (Fe_2O_3). White dots in (b) and (c) corresponds to atoms in lattice points. Red boxes show that there are two different atomic arrangements of iron oxide. [42-44]

(Due to the limited number of pages, only X-ray diffraction and microscopy are included here, but there are a lot of techniques that allow us to study the internal/surface structures and composition of materials such as neutron diffraction (crystal structure), scanning tunneling microscopy (atomic-scale surface structure), secondary ion mass spectroscopy (composition), and so forth. Materials scientists use all these characterization techniques to understand how atoms are arranged and how those atomic arrangements are related to material properties.)

A microscope can be said to be a materials scientist's best friend. Because the material properties vary depending on its composition and atomic arrangement, it is necessary to be able to find out information about the material's composition and atomic arrangement.

The state-of-the-art microscopes used by materials scientists help us observe microstructures of materials as well as atomic arrangements and even enable us to analyze their composition.

(blank page)

Ch. 5. Materials Thermodynamics

When we try to make materials, simply mixing random elements and wishing for good materials would take forever. If we're able to predict which materials are likely to form, we can create good materials much more efficiently. If elements are mixed under a given temperature and pressure, it is possible to guess what kind of materials will be made and what microstructures will be formed. This theoretical area of materials science is called "Materials Thermodynamics".

The most important concept in materials thermodynamics is the change of energy that occurs when making materials at a given temperature and pressure. If I want to make an atomic array of a material, materials thermodynamics theory allows me to calculate the change of total energy of the material during the process of creating said array. **According to the principle of materials thermodynamics, all materials want to take the atomic arrangement that has the lowest total energy**. Therefore, if we can find the lowest energy arrangement theoretically possible using material thermodynamics, we can predict what material will be formed. In other words, materials thermodynamics gives us the amazing power to predict which materials can be obtained without using experimentation.

Then, where is the energy of the material from? This is a slightly demanding topic but let us explain it carefully. There are three different types of energy in materials. The first is the energy associated with the bond between the atoms. The second is the energy associated with the degree of disorder of the atomic arrangement. The third is the energy related to the volume change at a given pressure. When a material is formed, the bonding between atoms changes, the degree of disorder of the atomic arrangement changes, and the volume changes. As a result, the total energy of the material changes. When we make a material, the atoms are constantly moving until they attain the atomic arrangement with the lowest possible energy.

Atomic bonding

According to quantum mechanics, atoms generally tend to be attached to each other. Thus, the atoms are always stuck together when the temperature decreases (in other words, when the thermal vibration is too small to break chemical bonds). When two atoms are chemically bonded, there is a minimum energy required to separate them at absolute zero temperature. This energy is called bond energy (E_b). For example, oxygen molecules are made of two oxygen atoms, and the minimum energy required to separate these atoms is referred to as the bond energy between oxygen atoms (**Figure 37**). By convention, the energy of an atom that is infinitely separated from its neighboring atom is zero. If these two

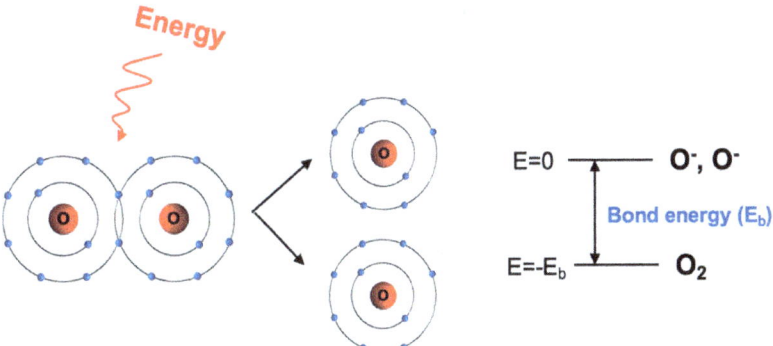

Figure 37. The minimum energy needed to remove the bond is called the bond energy (E_b), which is the positive quantity. By convention, we assume that the energy state when the two atoms are infinitely separated is zero. Then, when the atoms are combined, the energy value becomes negative ($-E_b$). That is, the larger the bond energy, the lower the energy state.

atoms are then bonded, the energy of the atom becomes lower by the magnitude of bond energy. Thus, the bonded system has a negative energy value (= $-E_b$).

The total negative sum of bond energy for all bonds available in materials is called internal energy (precisely speaking, the internal energy is the sum of the kinetic and potential energy of all atoms, but the total sum of bond energy is a good approximation for solids). According to the principle of materials thermodynamics, a material wants to be in the lowest energy state, so it tends to form an atomic arrangement that has the lowest internal energy. In other words, materials prefer to have chemical bonds with larger bond energies because having a greater

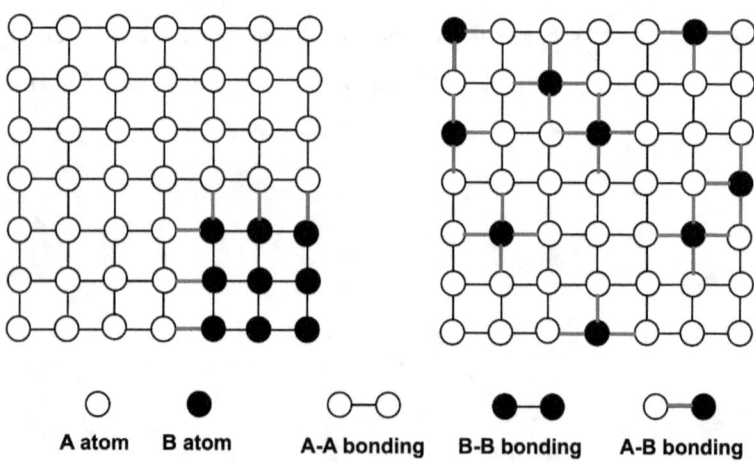

Figure 38. If A-B bonds have a much larger bond energy than the A-A bond and B-B bond, having more A-B bonds can have a much lower energy state (large negative energy). Thus, the material on the right is preferred (because the left structure has 6 A-B bonds, but the right structure has 29 A-B bonds.) On the contrary, if the A-B bond has a much smaller bond energy than A-A bond and B-B bond, having more A-B bond has a much higher energy state. Then, the materials will be separately rather than mixing A and B, so the material on the left will be preferred.

number of them lowers the internal energy (meaning the material has a larger negative energy value).

Let's discuss the role of bonds in atomic arrangement considering two different cases: (1) Elements A and B are completely separated, (2) Elements A and B are randomly mixed (**Figure 38**). Because the internal energy of solids is the total negative sum of the bond energy, the internal energy difference between these two cases appears as the difference in the total negative sum of bond energy.

First, let's assume that the A-A bond and the B-B bond have a smaller bond energy than the A-B bond. Remember that chemical bonding lowers the energy of materials. In this case, the material prefers to have A-B bonds rather than A-A or B-B bonds. This means that the A and B elements prefer to be mixed (the right image of **Figure 38**). Depending on the strength or directionality of the bond, mixing can be random or regular. The former is called solid solution, and the latter is called a compound (**Figure 39**). On the contrary, if the A-B bond has a

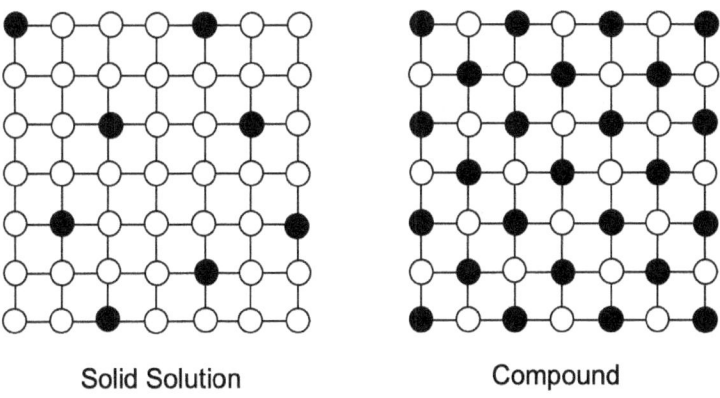

Solid Solution   Compound

Figure 39. If the second element is arranged in the solid, it is called a solid solution (left), and if all elements are arranged regularly, it is called compound (right).

 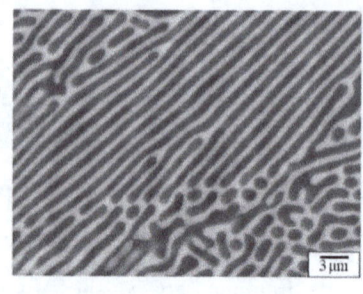

Figure 40. When water and oil are separated, the internal energy of the mixture can become lower, so it does not mix (left). In fact, when a solid is created, separation occurs for the same reason (right). Since water and oil are liquid, heavy water goes down and light oil goes up due to the gravity effect. In the case of solids, however, the gravity effect is weak, so the local separation occurs for a short time and the undulated microstructure is made. The picture on the right shows a microstructure separated by magnesium (Mg) (white) and C14-Mg2Ca (black).[45]

less bond energy than the A-A bond and the B-B bond, the material will prefer the separated state (the left image of **Figure 38**). In this case, A and B never mix.

The separation of water and oil can be explained by changes in internal energy (**Figure 40**). Water molecules have polarity, meaning they have a positive side and a negative side. Because of this, they are attracted to one another by strong Coulomb forces—the positively charged side of a water molecule is attracted to the negatively charged side of another. In other words, the bond between two water molecules has a high bond energy. Therefore, if a water molecule is bonded to another water molecule, water has a low internal energy. The bond between an oil molecule and a water molecule, on the other hand, is weak because an oil molecule does not have polarity. Thus, an oil-water bond does not decrease the internal energy as much as a water-water bond. Therefore, water and oil prefer to be separated. This is the case when the bond energy of the A-B bond is smaller than that that of the A-A bond and the B-B bond (the left image of **Figure 38**).

Entropy: Degree of disorder

Let's discuss the degree of disorder in an atomic arrangement and its relationship with the energy of materials. Let's assume that there is little difference in bond energy between the A-A bond, B-B bond, and A-B bond. In this case, there is no big difference in the internal energy of the material whether A and B are mixed or not. Based solely on internal energy, the material does not seem to have a preference between the mixed and the separated states, but is this really true?

If $E_{b,A-A} = E_{b,B-B} = E_{b,A-B}$ and $T \neq 0$ Kelvin

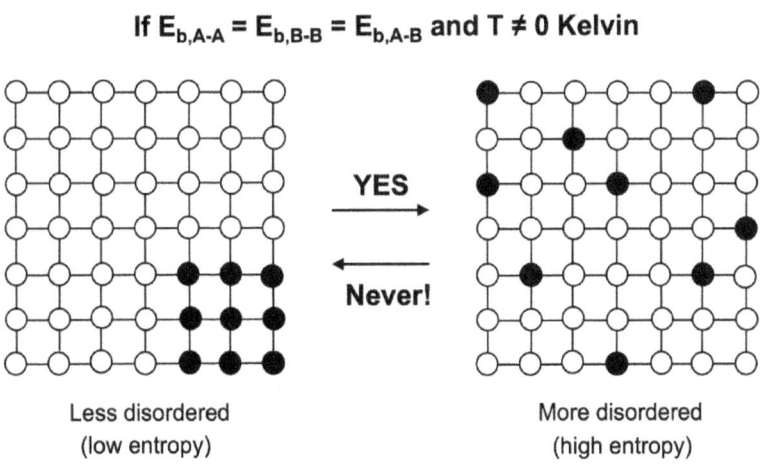

Less disordered
(low entropy)

More disordered
(high entropy)

Figure 41. Even though there is no change in internal energy by mixing, the material prefers to be mixed at a finite temperature. If the temperature is not absolute zero, the atoms can move slowly and eventually mix. In this case, materials thermodynamics says that the material prefers the state of high entropy (more disordered).

In this case, according to materials thermodynamics, mixing is preferred if the temperature is higher than absolute zero (**Figure 41**). The atoms always vibrate at a finite temperature above absolute zero. Then, at a certain point, the neighboring atoms are able to push an atom just enough to make it leave its original position. If this atomic motion occurs continuously at random locations, atoms can flow around (even in a solid state). This phenomenon is called diffusion. As a result, different elements are inevitably mixed at a finite temperature. Diffusion occurs more actively at higher temperatures, meaning that mixing occurs more rapidly. In summary, so long as atomic vibration occurs, a material mixes its constituent elements spontaneously.

An easy example is water and ink. If you drop ink in water, the ink spreads. Water and ink molecules are highly mobile in the liquid state and eventually mix (**Figure 42**). You have most likely never seen the ink flow into one corner spontaneously. If the atoms and molecules can move around, the tendency to mix always exists. If expressed in terms of materials thermodynamics, the material always wants to be more

Figure 42. When the ink is dropped in the water, the ink spreads naturally. We have never seen the ink reversed back to the corner of water cup. In other words, the material always has a tendency to be mixed (or disordered).[46]

disordered. Water and ink prefer to mix because the mixture of ink-water is much more structurally disordered than separated ink and water. Scientists have devised a quantity that describes the degree of disorder in a material. This quantity is called entropy. As the degree of disorder increases, so does the value of entropy. As you can see in the example of water and ink, a material always prefers to have a higher entropy—increased disorder.

The mixing of a material is greatly affected by the temperature. For example, if the temperature is absolute zero, the atoms do not move. Thus, mixing does not occur. However, if the temperature is higher than absolute zero, the atoms vibrate and move around, causing mixing to occur. For instance, salt is more soluble in hot water than in cold water. The higher the temperature, the more active the atoms. This means that there can be a significant increase in disorder. Therefore, according to materials thermodynamics, entropy and temperature are closely related.

Materials scientists found that the entropy effect on the energy of a system can be represented as a negative number—the multiplication of the entropy value (S) and the absolute temperature (T), $-S \times T$. Note that entropy has a unit, Joule/Kelvin and absolute temperature has a unit, Kelvin. Therefore, $-S \times T$ has a unit of Joule, which is the unit of energy (Joule/Kelvin $\times$ Kelvin = Joule). Now, let's consider that $-S \times T$ is the energy of the system related to the degree of disorder in the atomic arrangement. Based on materials thermodynamics, a material always prefers a lower energy state. This means that the material always prefers the atomic arrangement with the lower $-S \times T$ value. Thus, at any given temperature, the entropy value always wants to grow to make the $-S \times T$ value more negative (because T is always a positive number). For instance, at room temperature, ink and water mix to lower the $-S \times T$ value by increasing S.

Wait a second! If we go back to the example of a water-oil mixture, it seems that the separation of oil and water violates the idea of entropy because the water and oil don't mix (they do not want to increase the degree of disorder). What's wrong here? So far, I have explained two different energy contributions. The first is internal energy (E), and the second is entropy effect ($-S \times T$). These two contributions should be considered together allowing us to develop the following relation:

(total energy) = (internal energy effect) + (entropy effect)

$$E_{tot} = E - S \times T$$

So, it is necessary to find out which atomic arrangement provides the lowest total energy ($E - S \times T$).

In the case of water and oil, if we consider the entropy effect, the two want to be mixed. If we consider the internal energy (combined energy), they don't want to mix. In other words, the effect of entropy and the effects of internal energy compete. At room temperature, the bond energy between water molecules is very strong due to the polarity of the water molecules. Thus, the effect of the internal energy (E) is much greater than the effect of entropy ($-S \times T$). In this case, it is okay to ignore the entropy effect ($E_{tot} \approx E$), meaning the state of a water-oil mixture is determined primarily by the internal energy. Interestingly enough, if the temperature (T) increases, the $-S \times T$ term becomes increasingly dominant. Above a certain temperature, $-S \times T$ could be more dominant than E, so

$E_{tot} \approx -S \times T$, implying that it is possible to have a perfect mixture of oil and water at elevated temperatures!

Pressure effect

Finally, let's consider the effect of pressure on the energy of materials. When two different elements, A and B, are mixed—compared to when they exist separately—the volume is slightly different. For example, if the bond distance of the A-B bond is shorter than that of the A-A or B-B bond, the volume will decrease when the two materials are mixed (**Figure 43**). Based on materials thermodynamics, a material always tries to change its volume in the direction of pressure. Let consider the case that the bond distance of A-B is shorter than that of A-A or B-B. If we assume that the applied pressure compresses the material during mixing, the material will prefer to be mixed. This is because a decrease in volume is more natural under compression. Similarly, carbon prefers to

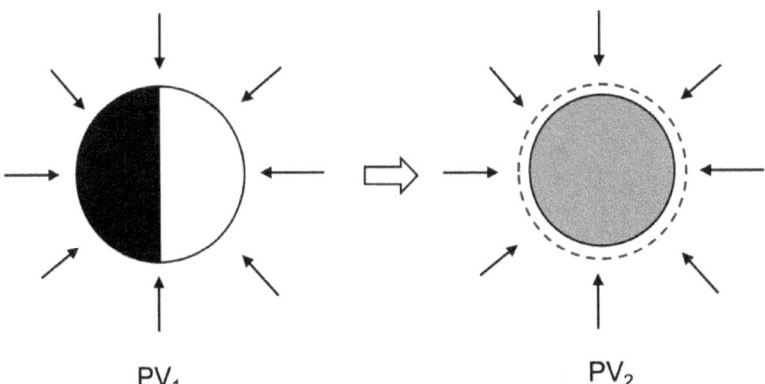

PV_1 PV_2

Figure 43. Under the pressure (P), if the volume changes by mixing materials, the energy state of the material changes by $PV_2 - PV_1 = P(V_2 - V_1)$. When the volume changes in a direction to adapt to the given pressure, the energy of the material decreases, and when it changes in the direction contrary to the pressure, the energy of the material increases.

become diamond rather than graphite under compression because diamonds have shorter carbon bonds than graphite. Energy values associated with pressure effects are represented by $\mathbf{P \times V}$, where $\mathbf{P}$ is pressure and $\mathbf{V}$ is volume. P has the unit of N/m^2, and volume has a unit of m^3. Thus, $\mathbf{P \times V}$ has a unit of Joule, which is the unit of energy ($N/m^2 \times m^3 = N \times m = $ Joule). Note that in the case of compression, the value of pressure is positive, and in the case of tension, it is negative.

Total energy

In summary, the total energy at a given temperature and pressure can be calculated as the sum of the internal energy ($\mathbf{E}$), the entropic energy ($\mathbf{-S \times T}$), and the energy related to the pressure ($\mathbf{P \times V}$).

$$E_{tot} \text{ (unit: Joule)} = E + P \times V - S \times T$$

Again, according to materials thermodynamics, all substances in the universe spontaneously transform from high to low internal energy. Similarly, when we hold a ball at a high point and let go, it falls to the floor. This can be interpreted as the ball wanting to reduce its potential energy— the ball moves spontaneously from a high to low potential energy position. As such, atoms in a material move spontaneously until they form the atomic arrangement with the smallest total energy. The atomic arrangement with the minimum total energy will be the final product that forms. That is, materials thermodynamics allows us to find materials (atomic arrangements) with the smallest possible energy at a given temperature and pressure. In addition, materials thermodynamics help us understand which materials are impossible to create. If the total energy of the material is too high, nature does not allow us to form the material. Therefore, materials thermodynamics makes it possible to understand the limitations of material development (**Figure 44**).

In fact, many materials science students struggle when studying materials thermodynamics. Seok-Woo often hears from undergraduate students that it is the hardest class to understand. He remembers having a difficult time because it was not easy to understand the many non-intuitive concepts (entropy, internal energy, etc.) required for thermodynamics using complex mathematical methods. However, materials thermodynamics is a very important subject because it makes material development predictable. This makes it one of the most important classes in the Department of Materials Science and Engineering.

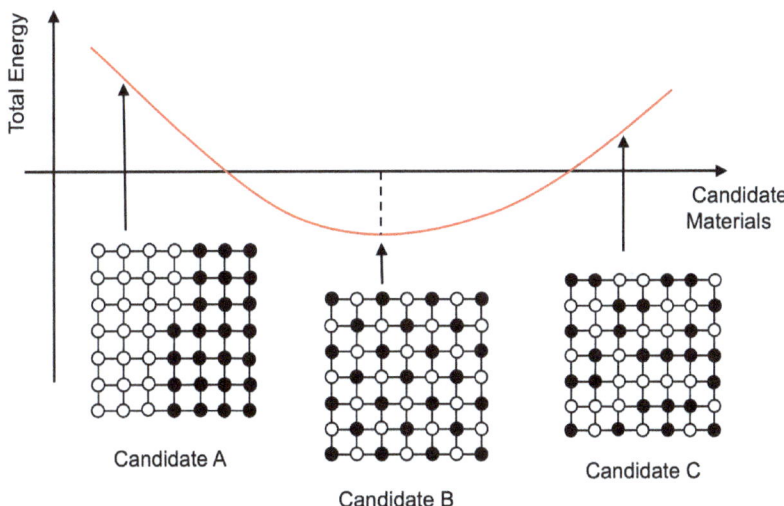

Figure 44. In consideration of all possible atomic arrangements, the material eventually takes the atomic arrangement with the lowest total energy state. In other words, the compound, which is a candidate B, is most likely to be formed. In this way, material thermodynamics can be used to know which atomic arrangements are most preferred or impossible to make.

Material thermodynamics is a study that theoretically explains what materials we can obtain and what materials cannot be obtained.

According to the theory of material thermodynamics, materials always want to have an atomic arrangement that gives the lowest total energy state under a given temperature and pressure, so it is very important to know how to calculate the total energy of the material.

Ch. 6 Phase Diagrams

Phase Diagrams and the Formation of Materials

While thermodynamics helps us understand why materials form, the key to knowing how they form lies in phase diagrams. Phase diagrams are among the most powerful tools in materials science, providing essential information about which phases exist under specific conditions of temperature, pressure, and composition. They serve as maps of stability, guiding scientists and engineers toward understanding how materials respond to changes in their environment.

Most people are familiar with the three classical phases of matter: solid, liquid, and gas. While these are indeed represented on phase diagrams, materials scientists must also account for the many distinct phases that can exist within solids themselves. For instance, metals can adopt different atomic arrangements depending on the conditions they experience. Iron, for example, can form either a face-centered cubic (FCC) phase or a body-centered cubic (BCC) phase, depending on which configuration minimizes the system's overall energy at a given composition, temperature, and pressure (**Figure 18**). The resulting atomic arrangement—governed by thermodynamics—is what ultimately determines a material's mechanical, electrical, optical, thermal, and magnetic properties.

The development of phase diagrams combines both theoretical modeling and experimental observation. Over the past century, materials scientists have conducted countless simulations, experiments, and thermodynamic assessments to chart the regions where specific phases appear. Modern computational tools now allow scientists to predict phase behavior even for complex, multi-component systems, accelerating the discovery of new materials.

If you wanted to modify the atomic arrangement of a material to improve its performance, consulting an existing phase diagram would be your first step. It would reveal how to tune temperature, pressure, or chemical composition to achieve the desired atomic arrangement. Whether designing high-strength alloys, semiconductors, or advanced ceramics, understanding and interpreting phase diagrams remains fundamental to controlling how materials form, transform, and ultimately perform in real-world applications.

Unary Systems: The Water Phase Diagram

A simple and familiar example of a phase diagram is that of water (H_2O), which maps temperature against pressure (**Figure 45**). You likely learned in school that water can exist as a liquid (what we drink), a solid (ice), or a gas (vapor). Although water can take on all three states, it typically occupies only one at a time. When we heat or cool water by placing it on a stove or in a freezer, we alter its environment, changing its phase.

The water phase diagram helps us predict these transitions. Its x-axis represents temperature, showing how thermal energy affects the state of matter, while its y-axis represents pressure, indicating how external pressure influences phase stability.

At a specific combination of temperature and pressure, we can determine which phase will exist. For instance, at 0 °C, water may exist as ice, liquid, or vapor, depending on the pressure. At 1 kPa, we are in the solid (ice) region, while at 1,000 kPa, we are in the liquid region. The diagram also shows that the boundary between solid and liquid phases has a negative slope—meaning that as pressure increases, ice melts. This happens because solid water (ice) is less dense than liquid water. When a skate blade presses down on ice, it increases pressure, locally melting the surface into a thin layer of water that acts as a lubricant (See the red arrow in **Figure 45**). This phenomenon allows a skater to glide smoothly across the ice with minimal friction.

The water phase diagram is an example of a unary system, meaning it contains only one component (H_2O). However, most real-world materials are composed of multiple elements. To understand those, we turn to binary phase diagrams.

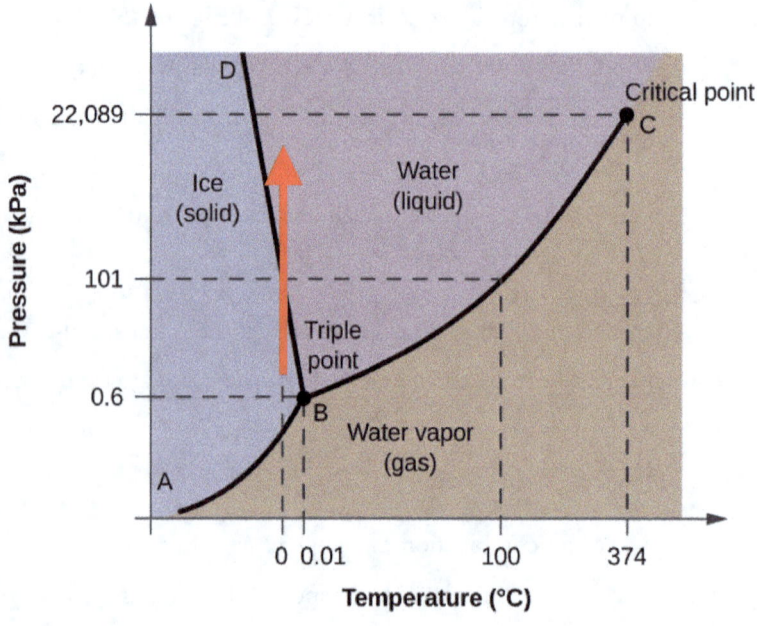

Figure 45. The labeled water phase diagram indicating the effect of temperature and pressure on phase present. Solid lines represent boundaries between phases. Dots represent key points. Different colored regions represent different phases.

Binary Systems: Two-Component Phase Diagrams

When a material contains two components, the number of possible phases, and therefore the complexity of its phase diagram, increases dramatically. A binary phase diagram represents how two components interact across varying compositions and temperatures (**Figure 46**).

Here, the x-axis represents composition, typically given in weight percent (wt.%) or atomic percent (at.%). For instance, a material containing 60 wt.% Tin (Sn) and 40 wt.% Bismuth has that ratio plotted along the x-axis. The y-axis represents temperature, reflecting how thermal energy influences which phases are present. Binary diagrams are typically constructed at a constant pressure (often 1 atm), so there is no pressure axis.

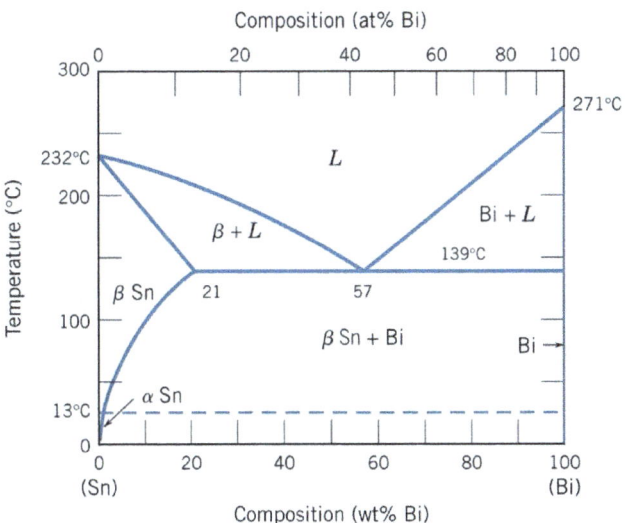

Figure 46. A binary Pb-Bi phase diagram showing how the phases present in a mixture of two elements changes with temperature.[23] It shows when the mixture is fully liquid, partly solid and liquid, or fully solid depending on the composition and temperature. In some regions, two phases can co-exist.

As the composition changes, even slightly, the resulting phases, and therefore the material's properties, can change dramatically. This is why phase diagrams are invaluable tools for materials design. They allow scientists to predict the effects of composition and temperature before performing costly experiments.

Think of a binary system like mixing chocolate and peanut butter. Depending on the ratio and temperature, you can have a fully mixed (liquid) state, a solid mixture, or regions where both solid and liquid coexist. Binary phase diagrams provide a map that tells us exactly what we'll get under different conditions—far more information than a unary (single-component) diagram can provide.

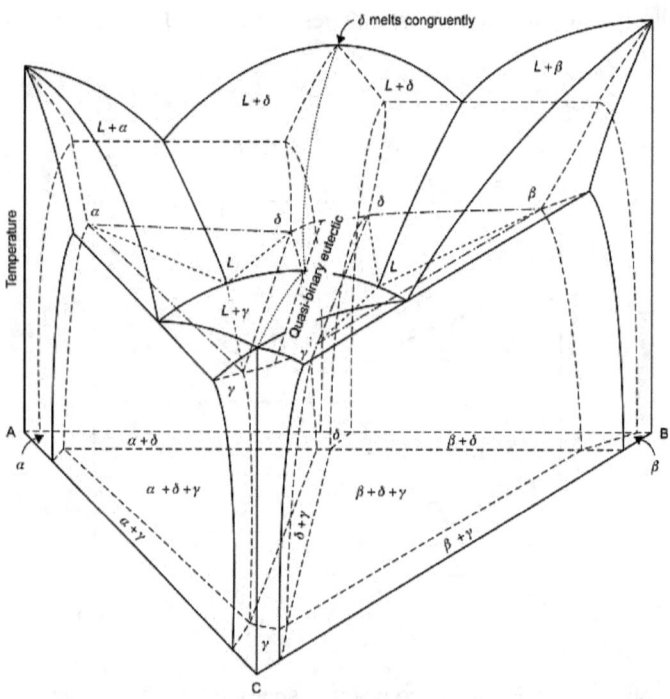

Figure 47. Example of three-dimensional phase diagram of ternary (A-B-C) system.[48]

As we move to ternary and multicomponent systems, phase diagrams become far more complex. Ternary diagrams (three components) are often shown in three dimensions or as slices, revealing new phase regions and reactions (**Figure 47**). For systems with four or more elements, visualization becomes impractical—so materials scientists use computational tools to predict phase stability across vast compositional spaces.

Case Study: The Iron–Carbon System

One of the most important binary phase diagrams in materials science is the iron–carbon (Fe–C) system (**Figure 48**). This diagram underpins the entire field of metallurgy because it describes how steels and cast irons (the most widely used structural materials in human history) form and transform.

The Fe–C diagram contains several distinct phases.

- Austenite (γ-iron) has a face-centered cubic (FCC) structure and exists over a wide temperature range between 0 wt.% and 2 wt.% carbon.

- Ferrite (α-iron) has a body-centered cubic (BCC) structure and can exist at room temperature.

Even though both consist of iron and carbon, their differing atomic arrangements give them distinct mechanical properties.

The diagram also includes an intermetallic compound, cementite (Fe_3C), which appears at a fixed composition of 6.67 wt.% carbon. Unlike other phases that exist over a range of compositions, intermetallics are

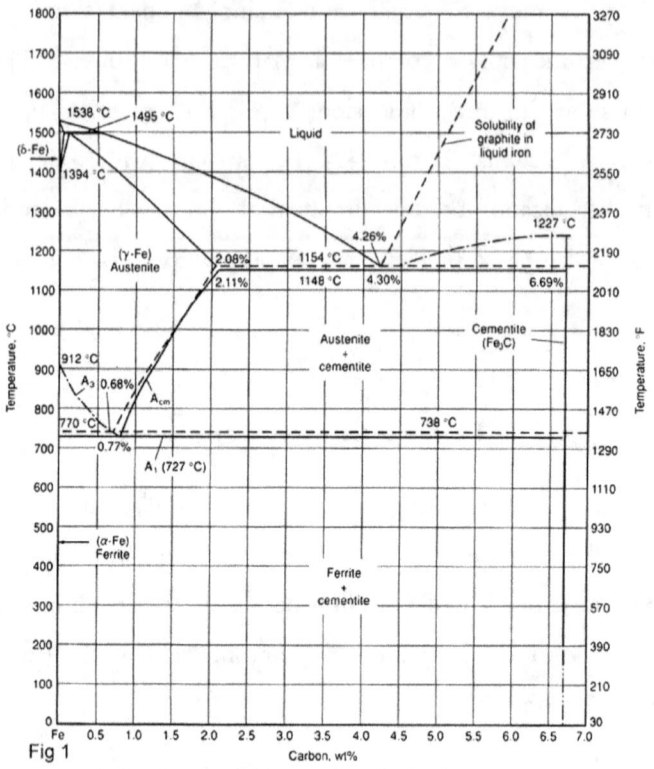

Figure 48. The iron-carbon equilibrium phase diagram with phases and key points (eutectoid, eutectic) labeled.[49]

stoichiometric, meaning they form only at specific ratios of elements. Cementite has an orthorhombic crystal structure and its mechanical properties very different from both austenite and ferrite.

In certain regions of the diagram, two phases can coexist. These two-phase mixtures form because the combined total energy of two phases is lower than that of a single phase. Phase boundaries mark where one phase transforms into another. The Fe–C diagram, with its multiple boundaries and transitions, is more complex than the simple water diagram, but is governed by the same thermodynamic principles.

84

The Power of Phase Diagrams

A deep understanding of phase diagrams is essential for designing and optimizing materials. These diagrams reveal what happens when different elements are combined, how compositions influence phases, and how temperature and pressure affect transformations. This knowledge enables engineers to tailor materials for strength, ductility, corrosion resistance, and countless other properties, while avoiding unwanted phases that could weaken performance.

Phase diagrams also guide manufacturing by indicating melting and solidification points, saving time and resources. Rather than relying on trial and error, materials scientists can consult a phase diagram to predict outcomes with precision. Whether designing a new aerospace alloy or a chocolate-peanut-butter blend, phase diagrams help us achieve the perfect combination of structure and properties.

Materials scientists devote immense effort to developing these diagrams because they allow us to predict how matter behaves and evolves. By combining theoretical simulations and experimental data, we gain insights into how nature forms materials, and how we can apply those same principles to create new ones. Through this understanding, we advance not only materials science, but the progress of civilization itself.

Phase diagrams use thermodynamics to describe which material phases are stable under specific conditions.

From unary phase diagrams, which explain the behavior of a single elemental material as temperature and pressure vary, to binary phase diagrams, which show what happens when two materials are mixed, phase diagrams allow us to predict which material phases will form under given conditions.

Ch. 7. Materials Kinetics

Materials thermodynamics tells us what materials we get but doesn't tell us anything about how long it takes to get them. If you can't create the materials that you want in your lifetime, is it even worth trying? There is a special meteorite made of magnetic material that has flown through the universe. The article about the meteorite states that people can't make this material because its atomic arrangement has formed slowly in the cold universe over millions of years. You probably don't want to wait for millions of years to make something!

When making new materials, there's usually a deadline. We can't just work forever! Thus, it is important to know how long it takes to get your desired atomic arrangement. Materials kinetics explains how quickly the atoms move around within a material. There are several important topics in materials kinetics. For instance, when a material is solidified in liquid, it is important to know the rate of solidification. It is also important to know how quickly elements diffuse into other materials when mixed in the solid phase at a specific temperature. This knowledge can then be applied to the process of improving materials by heating and cooling them.

Diffusion

Atoms in solids are easily spread at high temperatures. Temperature represents the degree of vibration in atoms or molecules. As was discussed earlier, atoms always vibrate if the temperature is higher than absolute zero. Vibrating atoms sometimes push their neighboring atom, which then leave their original positions and move to another location. As this phenomenon occurs over and over, the atoms in a material spread. The higher the temperature, the stronger the atomic

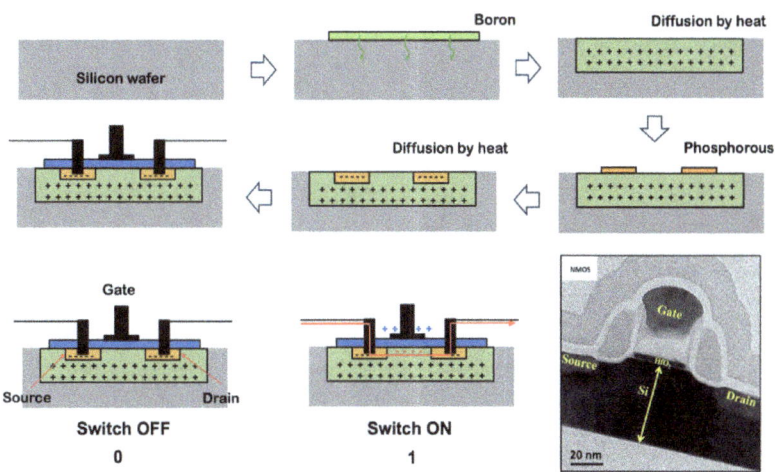

Figure 49. When boron is infiltrated into silicon, a semiconductor material, there is a lack of electrons, and when the phosphorus is infiltrated, surplus electrons are generated. Two negatively charged regions in the transistor structures are called source and drain. If the voltage is applied using the structure called the gate, it can open the way for electrons to move from the source to drain. In this way, the flow of the current can be turned on and off. This structure is called a transistor, and it is possible to perform logical operations with the binary calculation, 0 (current off) and 1 (current on).[50]

vibrations, the greater the spread. For metals and ceramics, the vibration of atoms at room temperature is not sufficient for diffusion to occur. Therefore, to move the atoms at a high rate, the material must be heated to an elevated temperature. Secondary elements can spontaneously penetrate a host material when the temperature is raised (i.e. the gear surface (**Figure 21**), or the semiconductor material (**Figure 49**)). If we know diffusion theory, it is possible to calculate the time it takes for elements to spread under a given temperature and pressure, and to adjust the spatial distribution of secondary elements accurately.

In general, atoms spread from high concentration to low concentration regions to increase the entropy of the system. The rate of

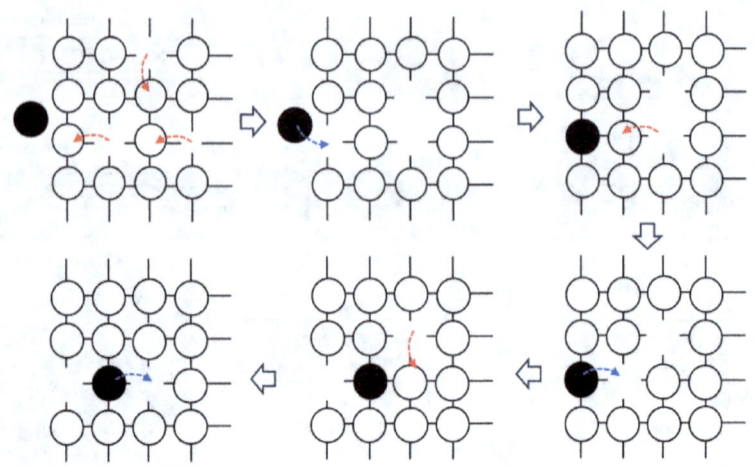

Figure 50. Diffusion mechanism using vacancy movement. The red arrow represents the vacancy movement, and the blue arrow represents the movement of the black atom. The size of the black atom is large and cannot diffuse easily through the lattice structure. Thus, the presence of a vacancy is essential for the diffusion of large atoms.

the atomic diffusion is determined by the weight of the atom, the temperature of the material, and the path by which the atom moves. If the secondary element is very small, such as carbon and hydrogen, it is easy for it to diffuse into the lattice structure of the host material. However, if the secondary atom has a similar size to the host atoms, it cannot easily spread. In this case, a vacancy must exist for the secondary atom to spread. It is important to know how many vacancies a host material can have at any given temperature and pressure. As shown in the figure, the secondary atoms can only spread when vacancies are present (**Figure 50**).

In fact, the existence of a vacancy can be easily explained by materials thermodynamics. Vacancies are created by breaking atomic bonds and removing an atom. This process increases the internal energy. Since materials want to have a lower internal energy by forming chemical bonds, the presence of vacancies is not preferred. However, the presence of vacancies allows the material structure to be more disordered. Thus, in terms of entropy, a material wants to have more vacancies. Then, the optimal concentration of vacancies under a given temperature and pressure is determined by the competition between internal energy and entropy. Above absolute zero, the material always has some vacancies, and as the temperature goes up, the number of vacancies does too (because the entropy effect becomes more dominant).

Solidification

Kinetic theory is also important when dealing with how quickly solidification occurs, both during the transition from liquid to solid, and when new materials are formed in solids. When liquid is solidified, the first step is the creation of a solid nucleus in several places. These nuclei grow and become solid (**Figure 51**). Since the growth of many nuclei occurs

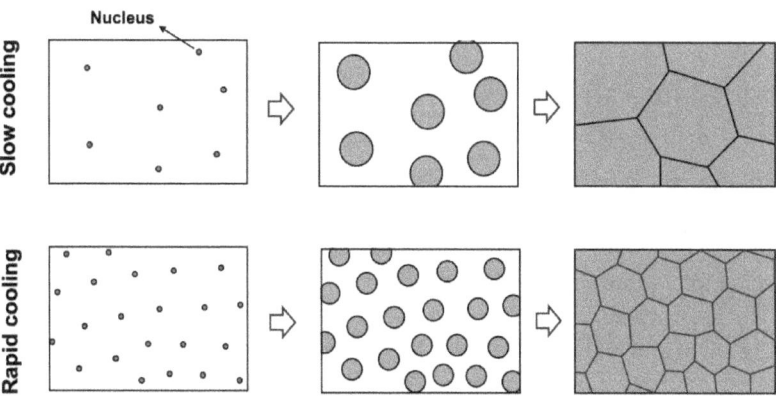

Figure 51. Slowly cooling the liquid material under the melting point of the melt results in a large grain structure, while the rapid cooling creates a small grain structure. If the liquid material is cooled extremely fast, it is possible to create a nano-scale grain structure. This is called a 'nanocrystalline' material, which exhibits superior mechanical strength.

91

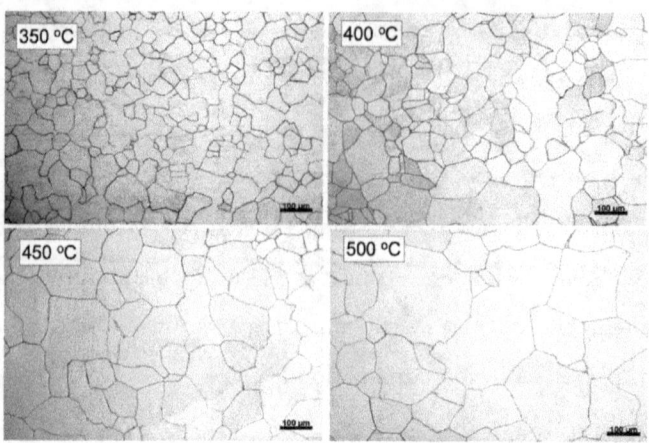

Figure 52. Changes in the grain size when the pure magnesium is heated for 120 minutes at the temperature written in the picture. With heat treatment at high temperatures, the movement of the atoms is more active, so the grains can grow faster, and the grain boundaries disappear. During the heat treatment at high temperatures, grains are combined, so only a few large grains remain. By controlling grain sizes, it is possible to control the properties of materials (e.g., smaller grains → harder).[51]

simultaneously, the material usually has multiple grains (this is termed a polycrystalline structure). Here, the size of the grains has a significant impact on the properties of the material. The size of the grains can be controlled by adjusting the temperature profile of the material at the time of solidification. For example, if the temperature of the liquid is rapidly lowered, many nuclei are formed in the liquid, which results in a fine grain structure (in other words, there are more grain boundaries in the material). Atoms in grain boundaries have increased separation compared to atoms inside of grains, meaning that many atoms in grain boundaries have broken bonds. This means that grain boundaries are regions with a higher internal energy. Therefore, thermodynamically, a material always wants to get rid of grain boundaries if the temperature is high enough to re-arrange the atoms present within grain boundaries. This phenomenon is called grain growth, which occurs at high temperatures (**Figure 52**).

Most Metals   Metallic Glass

Figure 53. The atoms in amorphous alloys are randomly arranged, like liquids. The bottom left picture is a fun LEGO situation that seems to protect a sample of amorphous alloy I made during the master's program. The lower right figure shows the irregular atomic array image taken using a transmission electron microscope.[52]

In some special metallic liquids, if the temperature drops rapidly, the atoms will freeze before they have time to form periodic atomic arrays (in other words, crystallization does not occur). In this case, the solidified material is a hard solid but maintains the random atomic array that was present in the liquid (**Figure 53**). When the alloy has the atomic arrangement of a liquid, it is termed an amorphous alloy—the strongest type of metallic alloy that humans can make. If we increase the temperature of an amorphous alloy, the atoms will eventually move around causing crystallization to occur.

Heat Treatment and Atomic Rearrangement

Because atoms can move more easily at higher temperatures, heat treatment—carefully controlling the temperature a material experiences—is one of the most effective ways to influence its internal structure. Understanding heat treatment and the underlying kinetics is essential for all materials scientists.

There are three fundamental stages in any heat treatment process: heating, holding, and cooling.

Heating: The first step is heating the material. As temperature increases, atoms gain kinetic energy and move more freely, allowing them to rearrange into lower energy arrangement. This can lead to phase transformations (as shown in phase diagrams) or to the reduction of crystal defects such as dislocations and grain boundaries. These structural changes can soften the material, relieve internal stress, and enable new grain structures to form. Different materials require different temperatures for effective heat treatment, depending on their melting point and composition.

Holding: Once the material reaches the desired temperature, it must be held there for a certain period (and no, don't hold it in your hands, hold it in a furnace!). Holding at high temperatures allows atoms to move, redistribute, and minimize internal stresses. If the holding time is too short, defects remain and desired transformations may not complete. If it is too long, unwanted grain growth may occur. The principle is much like baking brownies: take them out too soon, and they're undercooked, leave them too long, and they're burnt, but with the right time and temperature, you get the perfect result.

Cooling: After achieving the desired structure, the material must be cooled back to room temperature. The rate of cooling strongly affects the final microstructure and properties. Slow cooling gives atoms time to form stable, low-energy structures, producing softer, more ductile materials. Rapid cooling (such as quenching in water or oil) "freezes" the atoms in place, creating hard but often brittle structures.

To predict and control these transformations, engineers use a TTT (Time-Temperature-Transformation) diagram (**Figure 54**), which maps how cooling rate influences the resulting phases. By consulting the TTT curve, we can determine the specific cooling path needed to achieve a target microstructure.

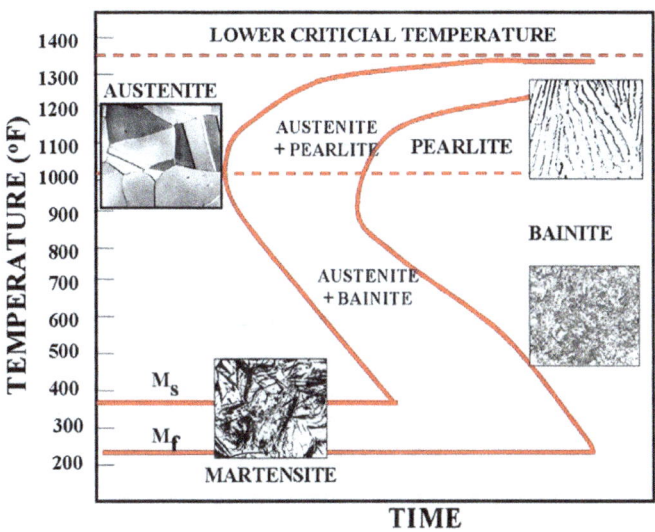

Figure 54. A TTT diagram of carbon steel showing the dependence of cooling rate on the microstructure formed. Letting the temperature drop slowly results in the formation of pearlite, whereas rapidly cooling results in martensite. Using a TTT diagram, materials scientists can predict which material they get and thus, control its properties.[53]

Figure 55. Steel bars during tempering process. The process was completed to with increased toughness and ductility.[54]

Common Heat Treatment Processes

A variety of heat treatment techniques combine these three stages in different ways to achieve specific material properties.

Annealing: In annealing, the material is heated to about 30–50% of its melting temperature, held at that level, and then cooled at a controlled rate. This process allows atoms to move freely, new grains to form, and dislocations to be eliminated, thereby reducing internal stresses. In the case of metals, the result is a softer and more ductile material—easier to bend or shape but not as strong. When a component is too hard or brittle, annealing helps restore its workability.

Quenching: Quenching involves rapidly cooling a material after heating. This sudden temperature change traps atoms in metastable arrangements, producing hard and strong—but often brittle—structures. The cooling medium (such as water, oil, or even unconventional materials like pig's blood in ancient sword-making) determines the rate of cooling and therefore, the resulting microstructure.

Tempering: Tempering is often performed after quenching to reduce brittleness and improve toughness (**Figure 55**). The material is reheated to a temperature below its lower critical temperature and held there, allowing limited atomic motion without changing the overall phase. Adjusting the tempering temperature and time enables fine control over strength and ductility—ensuring a balance between hardness and flexibility. If you want a metal bar that bends instead of breaking, tempering is the key.

Heat Treatment Beyond Metals

Heat treatment is essential not only for metals but also for amorphous materials such as glass. For instance, tempered glass, commonly used in windows and phone screens, is strengthened through controlled heating and rapid cooling. The outer layers solidify first and contract, while the inner layers remain slightly softer, creating internal compressive forces. Because cracking usually occurs under tensile forces, the internal compressive forces compensate the applied tensile forces, making the glass much more resistant to cracking or shattering than ordinary glass.

The Goal of Heat Treatment

The development and processing of materials can be thought of as a continual effort to achieve an optimized atomic arrangement that yields desired properties. Because atoms are too small to manipulate individually, we instead use thermodynamics and kinetics—through processes like heat treatment—to guide atoms toward configurations of lowest energy and greatest stability. In doing so, we transform raw materials into the advanced structures that support modern technology—from aircraft and bridges to smartphones and beyond.

Making materials involves changing the arrangement of atoms, so it is very important to know how fast the atoms can move and where they go. Materials kinetics allows us to predict the diffusion rate of atoms.

Because the diffusion rate of atoms changes depending on temperature, it is very important to properly control temperature (heat treatment) during the material manufacturing process.

(blank page)

Ch. 8. Material Properties

The primary purpose of developing new materials is to improve their properties. But what are the properties of materials? When a stimulus (mechanical force, electric field, magnetic field, light, heat, etc.) is applied to a material, that material exhibits a corresponding behavior. The quantified number of this behavior is called a material property. Material properties are values unique to a material and do not change with changes in size and shape. The easiest example is the comparison between density and mass. The mass of a material is proportional to the volume of the material. The density of a material, on the other hand, does not change alongside the size of the material. Because the volume and mass of the material increase proportionally, the ratio (mass divided by volume) never changes.

By changing the composition of elements and the arrangement of atoms, materials scientists create new materials, measure the material properties, and understand how to further improve these properties. Then, by understanding the relation between the atomic arrangement and the material properties, they can discover an optimal composition and atomic arrangement that yields the best material properties.

The most interesting part of materials science is the study of the relation between the atomic arrangement of a material its properties. Finding new relations through experimental and theoretical studies is incredibly exciting. We often think that God gave us atoms, and the role of the material scientist is to create new substances by combining them. It's really fun to make new materials—just like playing with LEGOs or building something in Minecraft.

Mechanical properties of materials

Mechanical properties are the values that describe the deformation or fracture behavior that occurs when a mechanical force is applied to a material. Mechanical force (**stimulus**) is applied to a material, and a resulting shape change (**behavior**) occurs (**Figure 56**). Let's say there are rubber bands and metallic wires with the same thickness and length. If we apply some small force to the rubber band and metallic wire, the rubber band will visibly stretch, whereas there will be no clear change in the size of the wire. Once the applied force is relaxed, both the rubber band and metallic wire will restore their original shape. This recoverable

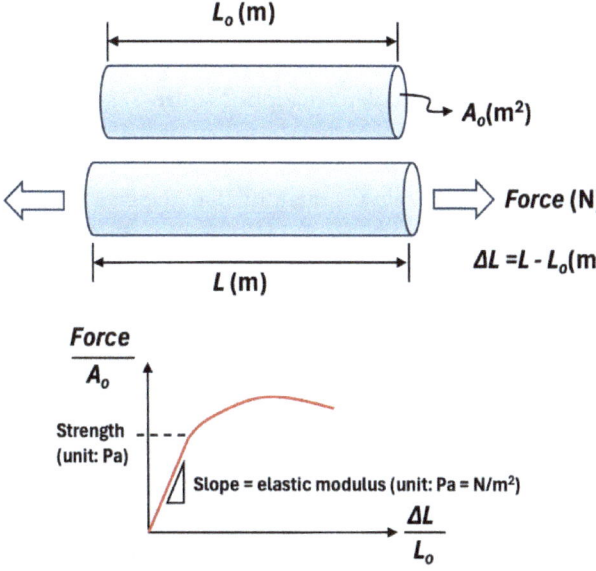

Figure 56. Tensile testing provides the mechanical data (red curve) of $\Delta L/L_0$ vs. (force)/A_o, where L_0 is the initial sample length, A_o is the initial cross-sectional area, and ΔL is the length change. When ΔL is small, the data show the linear line (elastic deformation), and its slope is elastic modulus. The (force)/A_o value at the moment when the curve begins to deviate from the linearity is strength of material (non-linear deformation corresponds to plastic deformation)

103

Figure 57. The elastic deformation of materials is widely used in sports. The football changes its shape when our foot hits it. Then, elastic recovery of the ball produces momentum causing it to fly away.[55]

deformation is called elastic deformation (**Figure 57**). As we all know, it is much easier to stretch a rubber band than a piece of metallic wire. The degree of elastic deformability is quantified as the elastic modulus. Since the metallic wire is much more difficult to deform than the rubber band, the metallic wire has a much higher elastic modulus than the rubber band.

The fundamental reason for elastic deformation is the desire of bonded atoms to maintain a constant separation (bond distance). If the bond distance is increased or decreased due to an applied force, when that applied force is relaxed, the atoms will always want to return to their original separation. As a result, materials can return to their original shape. Metals and ceramics have strong bonds between atoms, so it is not easy to reduce or increase the bond distance. Therefore, they have a high elastic modulus.

Figure 58. What happens if the wings of the plane suddenly bend in the middle of air and don't come back to their original shape?[56]

Another important mechanical property is the strength of the material. As mentioned above, if a small force is applied, elastic deformation occurs by slightly changing the distance between atoms. However, if the applied force is large enough to move atoms beyond a certain separation distance, the material cannot return to its original shape. This deformation is called plastic (or permanent) deformation. If the sample that we're testing is cylindrical, we can measure the critical force that causes plastic deformation by observing the point at which the material cannot fully recover. The strength of the material is then defined as the critical force divided by the sample's cross-section area. Knowing the strength of materials is very important in the design of products. Once we make a product, we want that product to keep its original shape for as long as possible. As an extreme example (**Figure 58**), imagine the shaking of airplane wings during flight. When the wings shake with the wind, the resulting bending must occur within the elastic deformation range. Imagine a scenario where the winds are incredibly strong. What if plastic deformation occurs and the wings don't return to their original shape? This is a dangerous situation and is why developing high-strength materials is incredibly important.

The higher the strength of the material, the more difficult it is for plastic deformation to occur. The armor worn by Iron Man must be made of ultra-high strength materials (**Figure 59**). To make this, he would need a special material that does not exhibit plastic deformation to withstand the many strong attacks that he faces. We haven't seen a material quite as strong as Iron Man's armor... yet.

Elastic modulus and strength aren't the only mechanical properties. Ductility is a measure of the stretchability of a material—how

Figure 59. Iron Man's armor is very strong and does not seem to be plastically deformed much (Iron Man, Paramount Pictures).[6,57]

much deformation can be withstood before the material is completely broken. This is usually measured as the percent length change before fracture. Toughness is a measure of how much mechanical energy per unit volume is needed to cause fracture. A material with a high fracture toughness can absorb a big impact before a crack begins to grow unstably and the material fractures. Hardness is a measure of a materials resistance to permanent deformation. A material that can withstand an indent or scratch will have a high hardness value, whereas a material that scratches or dents easily will have a low hardness and can be considered "soft." Materials scientists measure various mechanical properties and evaluate whether a material is mechanically stable under the forces it would experience in its intended application.

Studying the mechanical properties of materials has been Seok-Woo's main area of research for the past 20 years. To researchers in this area, a dream material would be high-strength, lightweight, and tough. Usually, strong materials are heavy—heavy iron is stronger than light plastic. This means that it is very difficult to find a material that is both light and strong. Iron Man's armor would require these types of materials. To prevent an attack, the material must be strong. To fly quickly, the material must be light. The development of these kinds of materials will have a tremendous economic effect. If the materials are strong, the products made from them can be used for a much longer time. Then, we do not need to as many products. This would lead to a reduction in the number of resources consumed. At the same time, if we develop light materials, cars and planes will be able to travel farther with the same fuel consumption, allowing us to save both fuel and money.

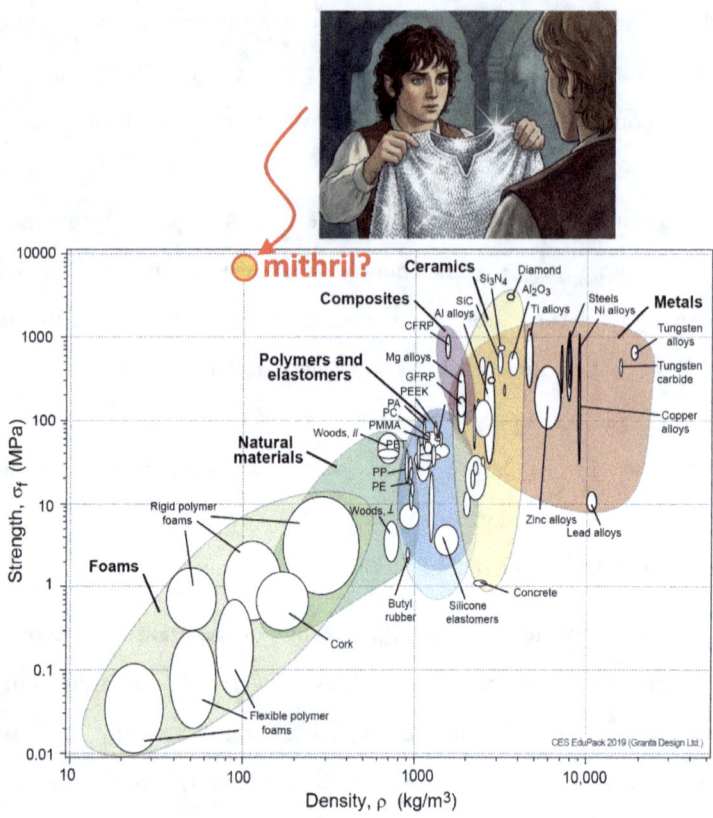

Figure 60. Mithril armor from the Lord of the Rings appears very light and super-strong so it plays an important role in protecting Frodo from enemy attacks. This Ashby graph shows the correlation between the strength and density of all the materials that humans have. Perhaps Mithril has a very high strength but low density. Of course, us humans don't have materials this light and strong. (**Load of Rings, New Line Cinema**) [58, 59]

Can we make these light ultra-high strength materials? Materials scientists have drawn graphs that plot the density vs. strength of materials. In general, light materials are weak, and heavy materials are strong. Therefore, making materials that are both light and strong is incredibly difficult. In other words, the strong cloth-based armor known as Mithril from the movie, the Lord of the Rings, would not be easy to make (**Figure**

60). However, material scientists are constantly studying and experimenting on various atomic arrangements that may reduce weight and increase strength simultaneously. If innovation occurs in this area of research, many of the things that you've seen in the movies will be realized.

Thus far, we've talked a lot about the theory of properties—stimulus and response. Though theory is important, the numbers that go alongside it are equally as critical. For example, yield strength is quantified by pascals, or newtons/meters^2. A material with a strength of 100 Pa is stronger than a material with a strength of 1 Pa. Thus, material properties are measurable quantities that become necessary for all sorts of applications. If a mechanical engineer wanted to build a car and needed it to be a certain strength, they would look at all material options and determine which are strong enough for their application. **Table 1** outlines the values of one such important material property (tensile strength) in several common materials used in high strength applications.

Material	Tensile Strength (MPa)
Iron	80–100
Titanium	100–225
Aluminum	5–20
Human hair	140–160
Bone	104–121
High-density polyethylene (HDPE)	26–33
Stainless steel	275
2800 Maraging steel	2,617

Table 1. Tensile strength of various materials

Electronic properties of materials

The electronic properties of materials quantitatively describe the behavior of materials when an electric field (**stimulus**) is applied. For

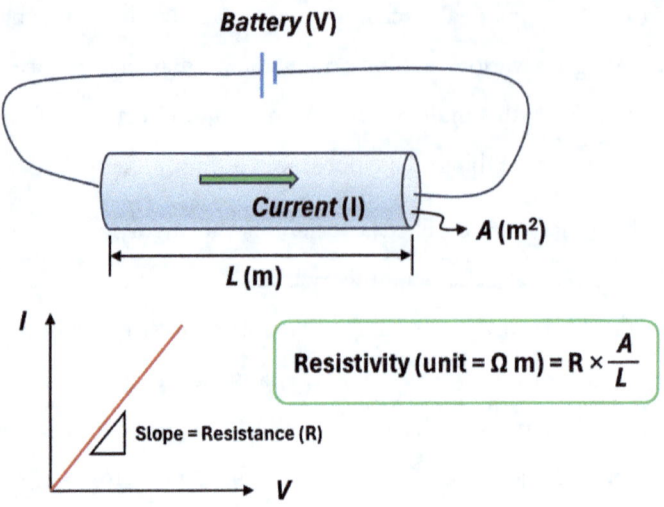

Figure 61. If the voltage is applied to a material, the electrical current flows. The resistance can be obtained from the I(current)-V(voltage) curve. Then, the resistivity can be computed if the dimensions of sample are known. Metals have the low resistivity, but ceramics and polymers have the high resistivity. Superconductor has the zero-resistivity material.

instance, under an electric field, electrons flow through materials (**behavior**). If we apply a voltage to the material (by connecting it to a battery), in some materials (conductors), the electric current will flow, while in other materials (insulators), the electric current will not flow. This resistance to the flow of electric current can be measured quantitatively by the value of resistivity (**Figure 61**). The lower the resistivity, the easier it is for electric current to flow through the material. In the case of metals, some electrons are not constrained by atoms. These electrons are called free electrons. If a voltage is applied, these free electrons move easily. Therefore, a metal has a very low resistivity (**Figure 62**). However, electrons in ceramics and plastics are bound tightly to the nucleus of the atoms, so even under a high applied voltage, the electrons do not move.

Figure 62. The copper or aluminum used for the transmission line has a high electrical conductivity. If the electrical conductivity is low, a lot of energy loss will occur through heat generation/dissipation during the current flowing, which is not good.[60]

Thus, insulators such as ceramics and plastics have a high resistivity, meaning that when a high voltage is applied, the electrons do not move.

Semiconductors fall between conductors and an insulators—the flow of electrons depends on the situation and can be adjusted. Using this property, it is possible to create a transistor that performs a binary calculation with 1 (when the electric current flows), or 0 (when the electric current doesn't flow) (**Figure 49**). The advancement of modern civilization can be represented by the development of computers, mobile phones, and displays, all of which are made using semiconductors. Without the development of semiconductor materials, we would still be living in the era of analog.

In relation to electronic properties, a dream material would be a room temperature superconductor. A superconductor is a material that has zero resistivity, meaning the electric current can flow without any loss of energy. Currently, all materials experience energy loss due to the

Figure 63. Unobtanium in the move Avatar (**20ᵗʰ Century Fox**) is known as a room-temperature ambient-pressure superconductor. Of course, this does not exist in our real world, yet.[17]

generation of heat as electric current flows (this is why your cell phone becomes hotter when it is used). If a superconductor could be used in electronic devices, there would be no energy loss and thus, increased efficiency. Perhaps those who have seen the movie Avatar remember that humans invade the place where the Na'vi tribe lives to take a mysterious floating material. According to a fan-made web site, this material is called Unobtanium, which is a room-temperature ambient-pressure superconductor (**Figure 63**).

In general, to obtain superconductivity, the temperature of a material must be very low. The best superconducting materials developed thus far require a near liquid nitrogen temperature (77K) for superconductivity to be observed at ambient pressure. Because of this, it is not easy to use superconductors in real life. If you develop a room-temperature ambient-pressure superconductor, you will be immediately

Figure 64. In 2023, physicists claimed that they discovered a room-temperature ambient-pressure superconductor, which is named LK-99. The recipe was relatively simple, so many research groups tried to make LK-99. Unfortunately, no one was able to reproduce superconducting phenomena from LK-99.[61]

granted the Nobel Prize in Physics. Many scientists have argued that they have developed a room-temperature ambient-pressure superconductor, but all of them turned out to be faux (**Figure 64**). We have often wondered if materials scientists will really make a room-temperature ambient-pressure superconductor anytime soon.

Material	Electrical Conductivity (σ(S/m))
Copper	5.96×10^7
Aluminum	3.5×10^7
Stainless steel	1.45×10^6
Germanium	2.17
Silicon	1.56×10^{-3}
Glass	10^{-15}–10^{-11}
Drywood	10^{-16}–10^{-14}
Niobium-titanium (at 9K)	∞

Table 2. Electrical conductivity of various materials

The most important measurable electrical quantity is electrical conductivity, which has units of σ, or Siemens/meter. If an electrical engineer wanted to create a robot, they would need materials with certain conductivities to create their desired circuits. Thus, materials scientists are needed to measure and determine the values of electrical conductivity in materials. **Table 2** outlines the values of electrical conductivity in several common materials.

Optical properties of materials

Different materials react to light (**stimulus**) differently. Some materials reflect light, while others transmit it (**behavior**). (**Figure 65**). When materials reflect a specific wavelength of visible light, we can observe various colors. In the case of transparent materials, the light passes through easily. The value that represents the degree of light penetration is transmittance. The higher the transmittance, the easier it is for light to pass through a material. The intensity of light before passing

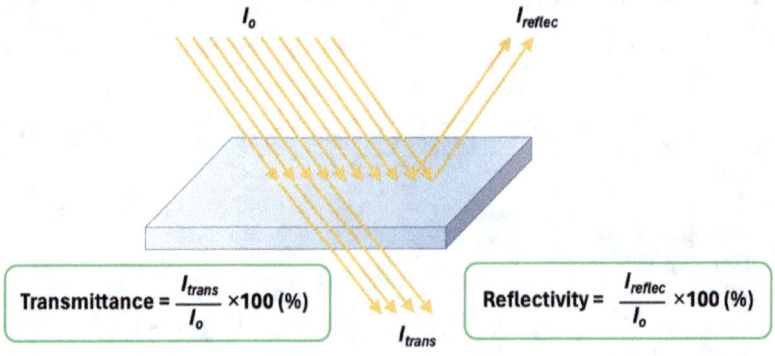

Figure 65. Materials reflect, transmit, and absorb light. If we know the intensity of light before it interacts with materials (I_o), the intensity of reflected light (I_{reflec}), and the intensity of transmitted light (I_{trans}), we can calculate what percentage of light is reflected, transmitted, and absorbed.

Figure 66. The primary mirrors of James Webb Space Telescope are coated with gold because gold reflects infrared rays very well. Infrared rays do not scatter with space dust, so they give us much clearer images of the universe. This is why the James Webb Space Telescope provides much clearer space images than the Hubble Space Telescope that uses visible light (Image from NASA).[62]

and the intensity of light after passing can be compared to determine the transmittance. A 100% transmittance material does not absorb the light at all, while a 0% transmittance material absorbs the light completely. The reflectivity (**Figure 66**), which represents the degree of light reflection, can be measured similarly by comparing the intensity of light before and after reflection. To increase the efficiency of solar energy, the reflectance of the solar panel must be reduced to improve light absorption.

Another optical property is refractive index. When light passes from one material to another, its path changes rapidly at the boundary of the materials. If you put chopsticks in a water cup, the chopsticks look bent. This is because the refraction of light occurs between the air and water.

The development of transparent electrodes is important when creating solar cells. A transparent electrode should transmit light well, and the electrical resistivity value should be low, properties which are difficult to obtain simultaneously. For example, light can easily pass through glass, but the electrical resistivity of glass is very high. In contrast, light cannot pass through metal, but the electrical resistivity is very low. We need to develop new materials that combine the advantages of glasses and metal, but this is a very challenging task.

To design a window, you need a material that transmits light. Using **Table 3**—which outlines the values of transmittance for several common materials—you can properly select the best material for your application.

Material	Transmittance
Glass (SiO_2)	0.84
Plexiglass	0.9
Quartz	0.9
Polyethylene	0.93
Aluminum	0
Silver	0
Copper	0
Anti-reflection coated glass	0.99

Table 3. Transmittance of various materials

Thermal properties of materials

Materials demonstrate certain thermal behaviors when they are heated or cooled (**stimulus**). If one end of a wire is heated, some materials would easily conduct heat to the other side of the wire, while others would not conduct heat at all (**behavior**) (**Figure 67**). In the case of metals, the vibrations of free electrons and atoms allow for the quick and easy conduction of heat. Materials such as Styrofoam, which do not have free electrons, do not conduct heat well. The value that represents the

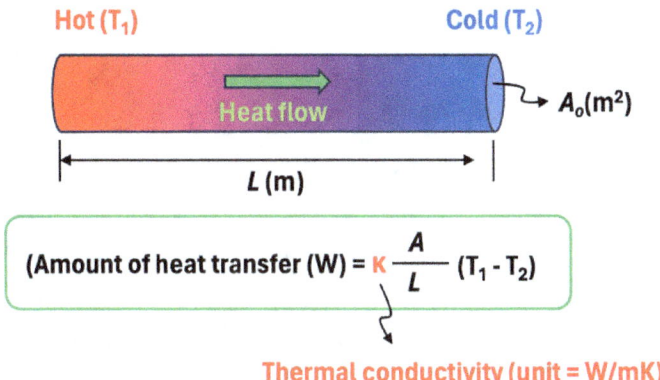

Figure 67. When the temperature gradient exists, the heat flows from the hot to the cold side. If we measure the amount of transferred heat energy per time (W) and know the temperature at both sides and sample dimensions, we can obtain the thermal conductivity.

conduction of heat in a material is called thermal conductivity. A material with high thermal conductivity can deliver the heat quickly, while a material with a low thermal conductivity cannot (**Figure 68**). For example, if you live in a town where the winter is cold and long, the house must be thermally insulated to prevent heat from escaping. It is necessary to use materials that have a low thermal conductivity so the loss of heat in the house is reduced and the cold from the outside cannot penetrate.

Air is an effective insulator that does not deliver heat well. Thus, it is possible to prevent thermal conduction by deliberately inserting air inside a material. An extreme example is a substance called Aerogel. Aerogel has a highly porous structure (**Figure 69**), meaning that most of the material (99.9% of its volume) consists of air, and thus, the material has an extremely low thermal conductivity.

117

Figure 68. The body of hyper-sonic plane must use the heat resistant materials to prevent the heat transfer into the plane. Maverick's hyper-sonic plane could not prevent the temperature increase and explodes in the air. (**Top Gun: Maverick, Paramount Pictures**)[63]

Another well-known thermal property is the coefficient of thermal expansion. Almost all materials increase in volume when heated. When heat is applied, the vibrations of atoms become more and more extreme, resulting in an increase in the separation between them. Since the degree of vibration of the atoms depends on the temperature, so does the volume of the material. The larger the coefficient of thermal expansion, the more the volume changes due to a temperature change. When joining materials at high temperatures, knowing how much the materials expand when heated is important because two materials could break at the joint if they swell or contract differently. Almost all materials naturally increase in volume when heated, but some special materials experience no change in volume. These special materials are called "zero thermal expansion materials". Charles Édouard Guillaume won the Nobel Prize in Physics in 1920 by discovering that the Nickel-Steel alloy does not expand when heated.

118

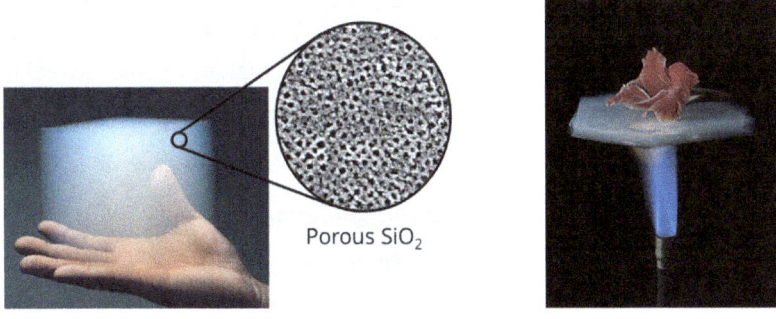

Porous SiO$_2$

Figure 69. Aerogel's 99.9% of the volume is filled with air. Thus, it has the extremely low thermal conductivity due to the lower thermal conductivity of air.[64]

If you tried to stir boiling water with a metal spoon, you'd notice that spoon heat up quickly. If you stirred that same boiling water with a wooden spoon, you wouldn't feel much of a change. This is thermal conductivity at work: how well the electrons move. Many applications require materials with specific thermal conductivities, whether that be insulation to trap heat inside of a house, or computer chips to get rid of any unwanted heat. Thermal conductivity is expressed in units of Watt per meter-Kelvin. The values of several common materials used in thermal applications are displayed in **Table 4**.

Material	Thermal Conductivity (W/(m×K))
Copper	342–413
Aluminum	220–240
Stainless steel	14.3–14.4
Germanium	59.9
Glass	1.05
Rubber	0.13–0.16
Polystyrene	0.03
Fiberglass (insulation)	0.04

Table 4. Thermal conductivity of various materials

Magnetic properties of materials

Magnetic properties refer to the values that describe the behavior of the material under an applied magnetic field (**stimulus**) (for example, when you bring a material close to a magnet). Some materials strengthen the magnetic fields around them by generating new magnetic field inside of them (**behavior**) (**Figure 70**). This phenomenon is called magnetization. The quantity indicating the degree of magnetization inside of materials is called magnetic susceptibility. Here, the ratio of the magnetic field generated inside of materials (**M**) to the external magnetic field (**H**) is the magnetic susceptibility (**M/H**). If the magnetic susceptibility is large, the material can be more strongly magnetized.

After magnetization, some materials maintain their magnetism even though the applied magnetic field is removed. These types of materials are ferromagnetic. For example, most of the magnets that you

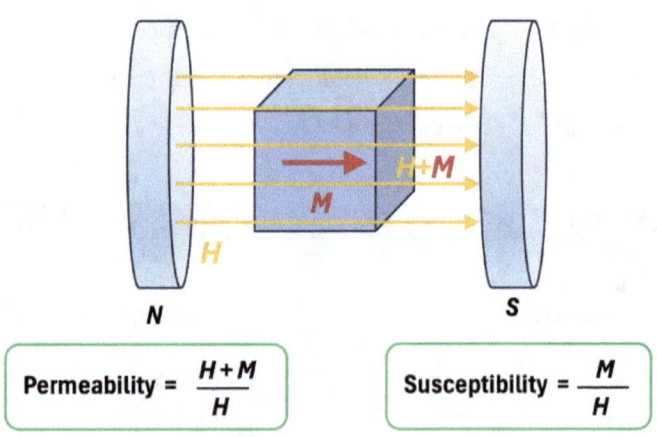

$$\text{Permeability} = \frac{H+M}{H}$$

$$\text{Susceptibility} = \frac{M}{H}$$

Figure 70. If the magnetic flux (H) is applied to a material, the material often generates additional magnetic flux (M) inside of it. By measuring either the total magnetic flux (H+M) or the magnetization of material (B), we can obtain the magnetic permeability and susceptibility.

might find on your fridge are made of an iron-based ferromagnetic material. Some materials, on the other hand, lose their magnetism immediately. This type of material is paramagnetic. The magnets that we usually use are, of course, ferromagnetic materials because they permanently maintain their magnetic properties.

Another well-known magnetic property is magnetic permeability. This quantity indicates how much total magnetic field is generated when a material is subject to a magnetic field. Both ferromagnetic and paramagnetic materials generate magnetic fields if an external field is applied. Thus, the total measured magnetic field increases. Here, the ratio of the measured magnetic field (B) to the external magnetic field (H) is the magnetic permeability (B/H). Surprisingly, there is a special type of material that can cancel out an external magnetic field. This material is called a diamagnetic material.

Superconductors are an example of a material that exhibits diamagnetism. They form a magnetic field equal to the applied external magnetic field. Since these opposite magnetic fields are repulsive, the superconductor can levitate above the magnet (**Figure 71**). This is called the Meissner effect. There are many applications that allow for the use of this levitation. One example is a magnetic levitation railway system. If a train were to be made of superconductors, that train would be able to levitate and move. Thus, there would be no friction between the railway and the train. This means that no energy would be lost due to friction, and in turn, the railway system would be incredibly energy efficient. The lack of friction would also mean that the train could move at a very high speed.

A permanent magnet is a material that can maintain a strong magnetic state. These highly magnetized materials can generate the

Figure 71. Internet meme related to the development of a superconductor. Here there is an exaggerated imagination that the city levitates in the sky.[65]

rotational force necessary for electric motors. The development of electric vehicles requires a lot of neodymium, a permanent magnet used in electric motors. Most of this material is buried in few countries across the globe (**Figure 72**). Thus, the US government has asked materials scientists to create new magnetic materials that have similar magnetic properties to neodymium.

As was discussed, one of the ways to describe the magnetic behavior of materials is magnetic susceptibility. A small and negative value means the material is diamagnetic: it is weakly repelled by the applied field. A small and positive value means the material is paramagnetic: it is weakly attracted to the applied field. A large and positive value means the material is ferromagnetic: it becomes magnetized under an applied field and retains its magnetism when that field is removed. **Table 5** outlines the magnetic susceptibility of materials commonly used in magnetic applications.

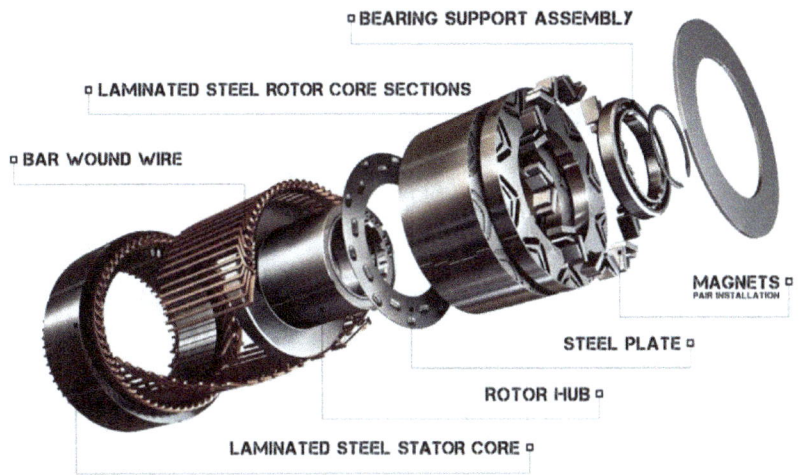

BEARING SUPPORT ASSEMBLY

LAMINATED STEEL ROTOR CORE SECTIONS

BAR WOUND WIRE

MAGNETS
PAIR INSTALLATION

STEEL PLATE

ROTOR HUB

LAMINATED STEEL STATOR CORE

Figure 72. The motor used for electric vehicles requires a permanent magnet. The reserves of rare earths used as a raw material for magnets are not sufficient, and the magnetic materials with strong magnetic force are being actively developed.[66]

As you've learned, one material can take on many different forms. A pure material will have dramatically different properties than an alloyed material. Materials that have undergone heat treatment processes will behave differently than materials that have not. A material will behave differently at hot temperatures than at cold temperatures. The tables given throughout this chapter highlight material property values at standard temperatures. To understand the properties of materials in specific conditions, find research studies that have measured those values rather than using standard tables.

Material	Magnetic Susceptibility
Iron	1000–5000
Nickel	600
Cobalt	250
Aluminum	2.2×10^{-5}
Stainless steel	10^{-5}–10^{-3}
Lead	-1.8×10^{-5}
Glass	-6×10^{-6}
Nickel-Iron	10^{4}–10^{5}

Table 5. Magnetic susceptibility of various materials

The properties of a material are the quantities that describe the behavior of the material when it is stimulated. Ultimately, a materials scientist is a person who creates materials with better properties.

To summarize, the properties of materials include mechanical properties, electrical properties, optical properties, thermal properties, and magnetic properties.

(blank page)

Ch. 9. Classes of materials

Metals

Metals are generally hard and glossy, and are characterized by high heat conductivity and electrical conductivity (**Figure 73**). An alloy consists of one or more metal elements (e.g., iron, aluminum, copper, titanium, gold, nickel) and often contains a relatively small amount of nonmetal elements (e.g., carbon, nitrogen, oxygen). Metals and alloys are almost always crystalline. In general, metals are denser than ceramics and polymers. Metals are typically strong, have high hardness, and are relatively ductile. They are widely used for structural purposes (reinforced bars for construction) because of their superior mechanical properties. Metallic materials have a large number of free electrons, making them an excellent conductor of electricity and heat. Optically, metals have a glossy appearance because visible light cannot pass through them. In addition, some metals (e.g., iron, cobalt, and nickel) have excellent magnetic properties.

Figure 73. Magnesium crystal[67]

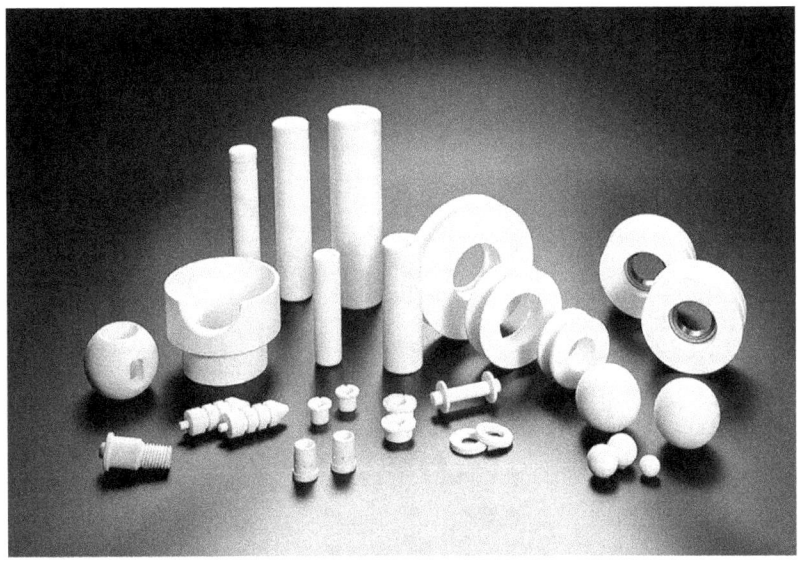

Figure 74. Ceramic materials are usually made by sintering powders at high temperature and under high pressure. Ceramics are strong and chemically stable.[68]

Ceramics

Ceramics are compounds of metallic and non-metallic elements and are characterized by excellent heat resistance and insulation, and electrical insulation (**Figure 74**). The atoms that make up ceramics are strongly bonded, making it difficult to break bonds, even at high temperatures. Thus, ceramics offer excellent heat resistance. In addition, the electrons in ceramics are strongly attached to the atomic nucleus, making them bad at conducting electricity. This electrical insulation is useful in electric circuits and devices when current needs to be blocked. Ceramics have excellent strength due to strong atomic bonding, giving them excellent wear resistance useful for the manufacturing of tools, bearings, and blades. The strong chemical bonds in ceramics do not react with external chemical attacks, making them excellent at resisting corrosion and oxidation. Ceramics are used in various areas such as the dishes we eat out of, furniture, electronic devices, auto parts, and tools. They are also attracting attention in applications such as artificial joints and sensors.

Semiconductors

Semiconductors are special materials where the electrical resistivity is higher than a conductor but lower than an insulator. When an external stimulus (heat, light, voltage, etc.) is applied, electrons can flow. We can control the conductivity of semiconductors by artificially doping (adding) secondary elements such as boron or phosphorous (see **Figure 49**). By using doping technology, it is possible to make an integrated circuit

Figure 75. The most amazing achievement of the development of human technology in the 20th century was the development of electronics such as computers and mobile phones using semiconductor materials like silicon.[69]

semiconductor material. The most common application of semiconductors is transistors, the basic components of modern electronic devices. Transistors play a role in the amplifying or switching of electric signals, which is essential for various electronic products such as computers, communication equipment, automotive electronic devices, and home appliances. Since almost all of the electronic products that we use contain semiconductor materials, it is no exaggeration to say that the development of human technology in the 20th century is in line with the development of semiconductor materials (**Figure 75**). In recent years, a lot of investment has been made in creating advanced semiconductor devices that may be used to improve the performance of artificial intelligence, computing, and robots.

Polymers

Polymers are materials made up of long chains of molecules, which form linear, branched, crosslinked, and network structures. The molecular structure has a significant impact on the physical properties of polymers. For example, linear structures allow for an increase in the flexibility of a material, while branched and network structures allow for an increase in strength. Polymers are generally used as an electrical insulator because they do not conduct electricity well. Many polymeric materials become viscous at a relatively low temperatures, making it easy to create various types of products (**Figure 76**). Polymers exist in various forms such as plastics, rubbers, and fibers, and are widely used in everyday life and industry. Most polymer materials are not easily decomposed once they are made and have been cited as the main cause of environmental pollution.

Figure 76. Plastics, one of the polymer materials, can be easily created by the forming process. The products made are light and strong, so they are also used for toys, packaging, and so forth.

In particular, the problem of marine pollution due to the presence of very small microplastics has emerged, and the development of polymeric materials that easily decompose is of urgent need.

Composites

Composite materials are made by mixing multiple different types of materials. In general, when we want to make up for the shortcomings of two different materials while combining their advantages, we create a composite material (**Figure 77**). For example, a ceramic is strong, but incredibly fragile. Metals are typically weaker than ceramics, but are much more ductile. Therefore, if the ceramic and metal are mixed, it is possible to create excellent structural materials that combine the strength of

132

ceramics and the ductility of metals. Composite materials developed using carbon fibers are both lightweight and strong. Thus, they are very widely used in automobiles and the aerospace industry. Recently, NASA sent a small helicopter to Mars, and the helicopter's propeller wing was made of a carbon fiber composite material, making it incredibly lightweight. The air density on Mars is only 1% of that of Earth's atmosphere. Due to the low air density, the propeller must rotate much faster, so the mechanical robustness of the propeller had to be good. This is why NASA chose to use carbon composites. Since composite materials combine dissimilar materials, it is essential to have good adhesion at the interface. If this adhesion is poor, a composite can be easily destroyed. Therefore, studying the interfacial bonding is one of the most important research topics in the field of composite materials.

Figure 77. The carbon fiber composite has a strong and tough carbon fiber in lightweight polymer matrix. Thus, it has high mechanical stability and lightweight. (Right) Lamborghini's frame made of carbon fiber composites. (Left) Ingenuity Helicopter on Mars. Its propellers were made out of carbon fiber composites.[70, 71]

Materials can be classified as metals, ceramics, semiconductor, polymers, and composites.

When an engineer makes a product, it is important to select materials based on their need. This is why the role of materials scientists is important during the manufacturing process because materials scientists know what the best materials are.

Ch. 10. Material Innovation

Computational materials science

As the performance of computers has improved, it has become possible to create hypothetical materials using computational programs. These models allow us to do things like predicting material properties, creating phase diagrams, or understanding materials kinetics. Molecular dynamics, an atomic-scale simulation technique used by materials scientists, can model the atomic arrangement of the material directly on the computer. This simulation makes it possible to describe how each atom moves because of changes in temperature and pressure. Molecular dynamics shows the motion of individual atoms that is difficult to see in the real world. Thus, it helps us understand the behavior of materials at the atomic scale through analysis of atomic movement (**Figure 78**). Another type of simulation is the first principle technique, which uses quantum mechanics and allows us to predict the atomic arrangement with

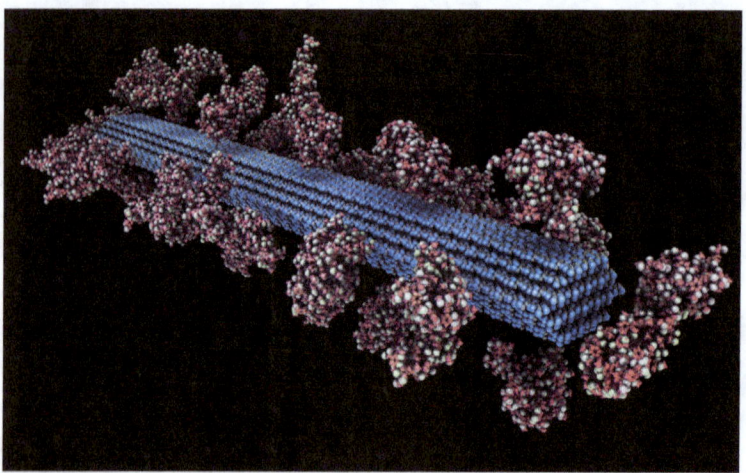

Figure 78. An atomistic model of cellulose (blue) surrounded by lignin molecules (green) comprising of a total of 3.3 million atoms (Bioenergy application). This phenomenon on the atomic scale is not easy to observe experimentation, so it is necessary to perform computer simulation to visualize what happens at the atomic scale.[72]

the lowest energy based on electronic structure calculations. If you like both computer programming and science, we strongly recommend you try computational materials science. This research collaboration allows scientists to make and test new materials on the computer, and then create those materials in real life using the simulation results.

Nanomaterials

If you cut a material, the total volume of the material remains the same, but new surfaces are created. This results in an increase in the total surface area. For any given volume, the smaller the size of the individual particles, the greater the surface area. A group of nanomaterials can have a very large surface area, making it possible to maximize the chemical reaction that occurs on the surface. There is much research to be done on

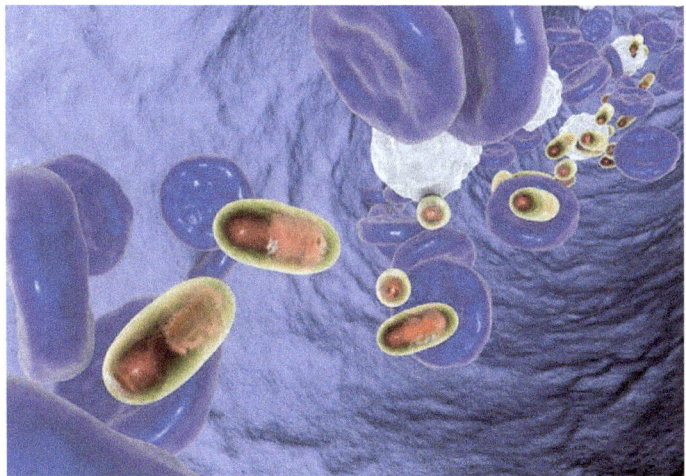

Figure 79. Nanotechnology is being developed to deliver medicine to the desired place by putting drugs in nanoparticles. It is a technology that can effectively heal without side effects by delivering drugs directly to cancer cells or sore areas.[73]

nanocatalysts, which can accelerate various useful chemical reactions (e.g., decomposition of carbon dioxide to prevent global warming). In addition, when a material becomes incredibly small (i.e. nano), it exhibits different properties compared to its bulk form. Materials scientists are making lots of efforts to obtain superior material properties from various nanomaterials such as nanoparticles, nanowires, nanotubes, and 2-D nanomaterials. Additionally, some special nanomaterials can be transferred into the human body and have the potential to improve the treatment of disease (**Figure 79**).

Biomaterials

Biomaterials refer to various substances used in the field of medical and biotechnology that interact with the human body. (**Figure 80**). Biomaterials must be safe for use in the human body. If we put a material in the body and it causes some sort of adverse reaction, that material is not biocompatible. Each biomaterial is selected and used for a specific application. For example, medical devices such as heart valves, artificial kidneys, cardiovascular devices, etc. can be used with highly bio-compatible polymer materials. A hip replacement is going to need to be made of a strong material, such as a metal or ceramic, that can support the weight of the body. In the field of tissue engineering, materials that promote interaction with cells and support tissue regeneration are utilized.

Biomaterials contribute to medical technology in modern medicine and life sciences and provide innovative solutions in fields such as organ transplantation, regenerative medicine, and dental treatment. The development of new biomaterials continuously progresses through clinical

Biomaterials

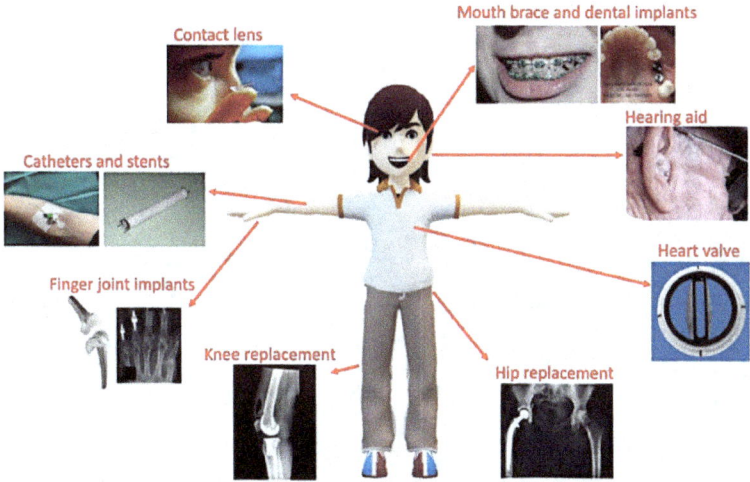

Figure 80. If some of the human body cannot function properly, it can be replaced with devices made out of biomaterials to sustain human life.[74]

trials and research, which can deliver more advanced treatment and healing technologies in the future. Undergraduate students studying biomaterials have created sutures that dissolve in the human body, gels that soothe the skin and prevent irritation, and so much more!

Eco-friendly materials

Eco-friendly materials minimize negative environmental impacts and emphasize sustainability. These materials are based on resources derived from nature or recycled. Representative eco-friendly materials include biomass-based materials, biodegradable polymers, and recycled materials. Biomass-based materials are renewable organic materials that

Figure 81. The development of new plastics, which can be decomposed faster, may reduce environmental pollution.[75]

come from plants and animals. Their use reduces dependence on fossil fuels and reduces greenhouse gas emissions. Biodegradable polymers are used as an alternative to plastics and are quickly decomposed under certain environmental conditions and thus, do not harm nature. Recycled materials are made into new products by recycling existing consumer goods, contributing to a reduction in resource consumption while simultaneously reducing the amount of waste (think about all of the cans and bottles you've returned!). Eco-friendly materials are applied in various industries such as architecture, fashion and packaging. They contribute to the preservation of our environment and its resources and promote the transition to a more sustainable future (**Figure 81**).

Energy materials

Energy materials are used in various technical areas such as energy production, storage and transmission. These materials play a key role in solar, wind, and power storage technology. The most commonly used material in solar panels is silicon. It plays a key role in converting energy from the sun into electricity. In wind power technology, turbine wings must be made of lightweight and strong composite materials to efficiently collect wind energy. In power storage technology, high-performance and high-efficiency energy storage materials like lithium-ion batteries are incredibly important. Since the Industrial Revolution, the concentration of carbon dioxide in our environment has increased due to coal, oil, and gas use, causing global warming. As a result, energy materials

Figure 82. Development of energy materials that can make energy generation and storage more efficient is essential for sustainable energy supply in the future.[76]

play an important role in mitigating the effects of global warming by reducing the use of coal and oil. The development of new energy materials for future powerplants, power grids, and batteries, is only possible through research and innovation. This is an important factor in promoting the development of an environmentally friendly and sustainable energy supply and responding to the pressing global challenges of climate change and energy security (**Figure 82**).

Additive Manufacturing

Metal additive manufacturing is a high-tech manufacturing technology which deposits metallic material layer-by-layer, forming 3D shapes and products (**Figure 83**). This technology makes it easier to produce objects with complex forms that existing metal processes cannot create. Metal additive manufacturing uses a laser or electron beam to form a metallic layer by melting metal powders. Three-dimensional printing of various metal materials such as steel, aluminum and titanium are possible. The biggest advantage of metal 3D printing is that the product that is formed is close to the shape of the desired final product, minimizing the unnecessary waste of materials caused by cutting and griding. Complex shapes can be manufactured with only one print, reducing the number of steps required to form the final product, whether that be welding or assembly. This simplifies the manufacturing process, improves production efficiency, and reduces manufacturing costs. During 3D printing, the metal powder melts and solidifies rapidly, meaning the atomic arrangements formed are drastically different from existing metal processes like casting, which undergo slow solidification. Therefore, metal additive manufacturing provides many new research opportunities, as we

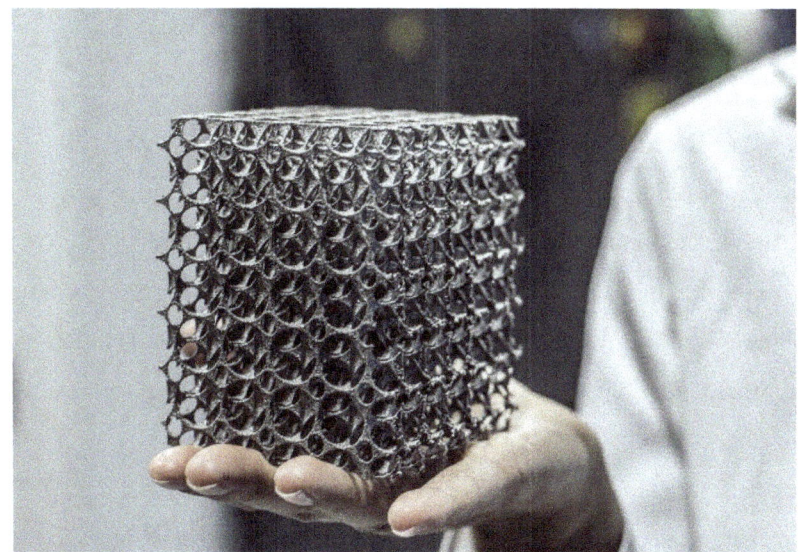

Figure 83. The metal 3D printing technology helps you easily create a complex metal structure that cannot be made in a conventional way. However, there are many cases where mechanical properties are often poor, so it is still a field that requires extensive materials science research.[77]

must study the relationship between the atomic arrangement and properties of 3D printed metals.

Quantum Materials

Quantum materials are substances that exhibit remarkable properties due to quantum mechanical effects at microscopic scales. These materials display phenomena such as superconductivity—the conduction of electricity without resistance at extremely low temperatures—and can act as topological insulators, conducting electricity on their surface while remaining insulating in their bulk. The unique behavior of quantum materials is governed by quantum mechanics. Principles such as wave-particle duality and quantum entanglement lead to a range of unusual and

143

often counterintuitive properties that are not observed in classical materials.

The applications of quantum materials are diverse and groundbreaking, spanning various fields of technology and science. In quantum computing (**Figure 84**), these materials are necessary to develop qubits, the fundamental units of information that leverage quantum superposition and entanglement to perform complex computations at unprecedented speeds. Additionally, quantum materials are integral to advancements in energy technologies, such as the creation of more efficient superconducting magnets for magnetic resonance imaging (MRI) and particle accelerators, as well as the improvement of energy storage and transmission through high-performance superconductors. They also enhance sensor technology, enabling extremely sensitive measurements for applications ranging from telecommunications to fundamental physics research.

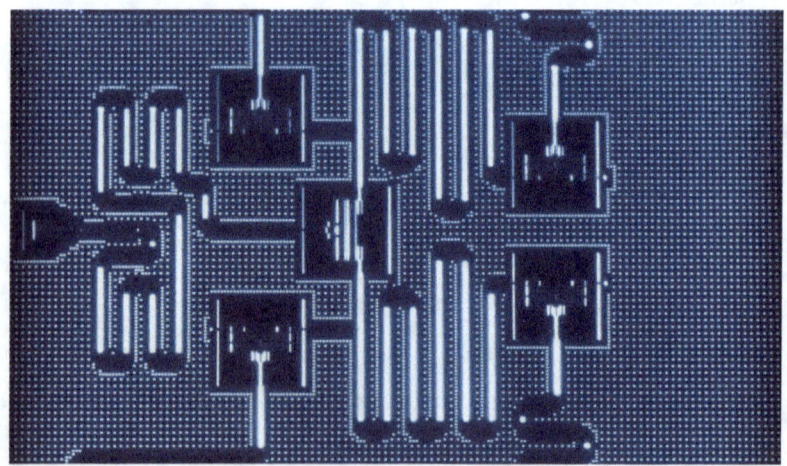

Figure 84. IBM's five qubit (quantum bit) process of the quantum computer, which is made out of superconductors. Now, IBM allows the public to access their quantum computers and to experience them.[78]

There are many exciting research opportunities in the field of materials science such as computational materials science, nanomaterials, biomaterials, eco-friendly materials, energy materials, metal additive manufacture.

Research in these area will play a key role in solving many critical problems related to global warming, climate change, energy crisis, and human health.

(blank page)

Ch. 11. Do It Yourself – Bubble Crystals

Creation of 2D bubble crystals

As described in **Chapter 3**, a crystal consists of atoms arranged in a periodic, repeating pattern. Interestingly, you can create a two-dimensional analogue of a crystal using bubbles floating on water. This bubble-crystal model not only makes it possible to visualize how atoms organize into a lattice, but also lets you deliberately introduce various imperfections such as vacancies, solute atoms, dislocations, and grain boundaries. Even more exciting, you can perform simple mechanical tests on this model. You can observe how plastic deformation occurs through dislocation motion and how fracture initiates along grain boundaries - phenomena that are often difficult to see directly in real materials.

All of these demonstrations are surprisingly accessible if you have the following items.

- Water
- Dish soap
- Container
- Fish tank bubbler
- Rubber hose
- Syringe needle

The key component in this experiment is the fish-tank bubbler. To form a proper bubble crystal, all bubbles must be the same size (typically a few millimeters in diameter). Because a fish-tank bubbler supplies a steady, consistent airflow, it allows you to generate bubbles of uniform size continuously.

First, fill a container with water and mix in dish soap. The required amount of soap varies by brand, so simply add more until the bubbles form easily and do not pop too quickly.

Second, assemble the bubbler by connecting the bubbler, rubber hose, and syringe needle (**Figure 85**). Smaller bubbles are generally better for forming a crystal, so a syringe needle with a small opening works best. If the needle produces bubbles that are too large, you can gently crimp the needle tip with a tool to reduce its diameter. If the bubbler's airflow is too strong, you can also create a few tiny holes in the rubber hose to let some air escape; this will reduce the flow rate at the syringe needle. (Note: The syringe needle is optional. If you don't have one, you can simply create bubbles directly from the rubber hose. For example, you can seal the end of the hose and make a tiny hole on its side; bubbles will then form at that opening.)

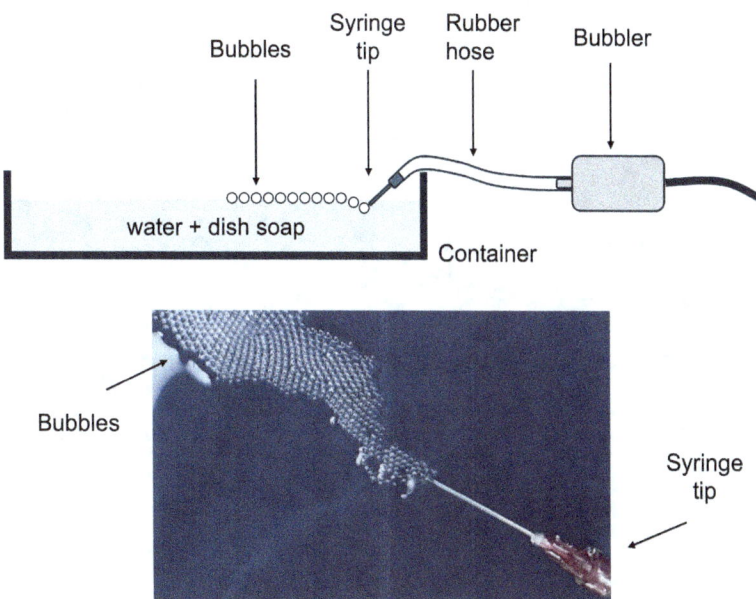

Figure 85. Schematic diagram of the bubble crystal maker. The bottom photo shows the formation of bubbles.

Third, turn on the bubbler and place the syringe needle into the soapy water. You should quickly see uniformly sized bubbles forming. If the bubbles rise too quickly and begin to pile up, gently blow them aside with your breath (or use a hair dryer on a low setting or portable fan). Soon, you'll observe the bubbles attaching to one another and arranging themselves into a periodic pattern (**Figure 86**). This attraction occurs because of surface tension, which pulls bubbles together—much like how atoms bond due to chemical forces (see **Chapter 5**).

Figure 86. Bubble crystal. The background appears blue because a blue container was used.

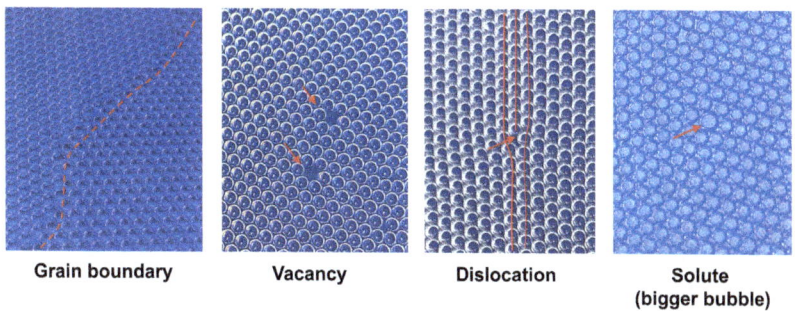

| Grain boundary | Vacancy | Dislocation | Solute (bigger bubble) |

Figure 87. Microstructure in a bubble crystal: grain boundary, vacancy, dislocation and solute.

Microstructure in a bubble crystal

It is possible to create various imperfections in bubble crystals such as grain boundaries, vacancies, dislocations, and solutes (**Figure 87**). Usually, when bubbles gather on the surface of the water, several small crystals form simultaneously and become attached to one another. Because these crystals grow in different orientations, grain boundaries develop between them. In addition, regions not occupied by bubbles remain vacant, leading to the formation of vacancies.

Even though the bubbler's airflow rate is constant, slightly larger or smaller bubbles still form occasionally. These bubbles act as solutes. During crystal growth, dislocations, which have an extra half atomic plane, also develop. They can be difficult to detect at first glance, but become much easier to observe once the crystal is deformed (**Figure 87**).

(****Please revisit Chapter 3** and compare real microstructures (**Figures 19, 20, 22, 24**) with those in the bubble crystal—they are remarkably similar!)

Deformation and fracture of bubble crystal

Once you have created the bubble crystals, it's time to conduct deformation and fracture experiments. You can use your hands or two rulers to compress or stretch the bubble crystal. To pull the crystal apart, simply place your hands or rulers on both sides of the crystal. Because of surface tension, the bubbles will adhere to them, allowing you to gently pull the crystal outward.

First, when you lightly compress or stretch the bubble crystal, you'll notice that certain features begin to move. These are not bubbles, they are dislocations (**Figure 88**). This clearly demonstrates that plastic deformation in bubble crystals occurs through dislocation motion, just as it does in metals (see also **Figures 22** and **23**).

Second, if you continue pulling the bubble crystal further, you'll begin to see fracture occur along the grain boundaries—the weak points in a crystal (**Figure 88**). A similar phenomenon occurs in ceramics, where fracture often initiates at grain boundaries.

Seok-Woo regularly uses bubble crystals in his classes. Because atoms are far too small to observe directly, it is challenging to show their behavior in a classroom setting. However, since bubbles behave in the same way as atoms, a bubble crystal serves as an excellent model for demonstrating atomic bonding, microstructure formation, and the processes of deformation and fracture. Seok-Woo hopes that readers of this book will try making a bubble crystal themselves and enjoy discovering the fascinating world of materials science!

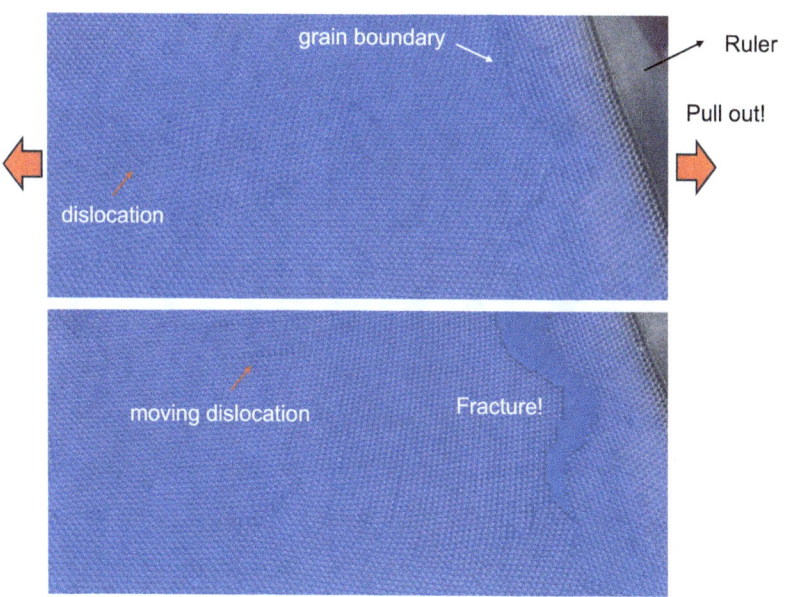

Figure 88. During deformation, dislocations move throughout the crystal. Small visible features within the crystals—one example indicated by the red arrow—are dislocations. If deformation continues, fracture occurs at the grain boundaries, which are the weak points in a bubble crystal.

Concluding Remarks

As technology develops, the performance of products will become increasingly dependent on the quality of the materials used to make them. In the past, people would use the materials that existed in nature with little to no modifications. Today, material scientists design and make new materials and use them to create high performance products. I strongly believe that in the near future, there will be an increased dependence on new materials, meaning materials science will become increasingly important for technological development. We are currently facing many serious problems. Fossil fuels will soon be depleted, and climate change and environmental pollution threaten our lives. The performance of airplanes and computers has already reached the limit of human technology. To solve these problems, technological innovation is essential, which must be preceded by the development of new materials. In the future, more material scientists will be needed, and the value of material science is expected to increase dramatically. I hope that all the middle and high school students who read this book become interested in the fascinating field of material science!

Seok-Woo Lee

Seok-Woo Lee

As a materials science and engineering student, I've learned a lot about materials. How to understand their structure, evaluate their properties and performance, and use this knowledge to make them better. I've also learned about how important materials scientists are. Students like you and I are going to be the ones that make room temperature superconductors, the materials for moon-habitats, Iron Mans armor, and so much more! The future of technology is dependent on us. Even though mechanical engineers make the cars, civil engineers make the bridges, and biomedical engineers make the implants, none of their jobs would be possible without materials scientists. For hundreds of years, mankind has explored the relationship between the structure and properties of materials. Because of this, you can pin a calendar to your fridge, fly over oceans, and watch movies on your iPhone. Just think about all the amazing things that we have yet to do because we don't have the right materials. With your newfound knowledge, step into the unknown, make something new, and explore. Who knows… you could be the one to make the materials that change the world!

Wyeth Haddock

Glossary and Index

Alloy. p.37   a mixture of two or more elements, typically metals, where the primary component is usually a metal, designed to enhance certain properties such as strength, durability, or resistance to corrosion.

Amorphous. p.31, 43   an atomic arrangement that lacks a defined, orderly structure at the atomic or molecular level, resulting in a non-crystalline and often irregular appearance.

Bravais lattice. p.32   a repeating, infinite arrangement of points in space that forms the foundational grid for describing the periodic structure of a crystal, where each point in the lattice has an identical environment.

Composite. p.42, 132   a material made from two or more distinct substances combined to achieve properties that are superior to those of the individual components.

Compound. p.66   a substance formed when two or more different elements chemically combine in fixed proportions, resulting in a material with properties distinct from those of its constituent elements.

Crystalline. p.31   the orderly and repeating arrangement of atoms or molecules within a crystal, which forms a well-defined geometric pattern extending in three dimensions.

Diffusion. p.88   the process by which particles spread from areas of higher concentration to areas of lower concentration, driven by their random motion.

Dislocation. p.39   a type of defect (one dimensional defect) where there is a misalignment in the regular arrangement of atoms, disrupting the crystal lattice and affecting the material's mechanical properties.

Electronic properties. p.109   The electronic properties of materials refer to how a material's structure influences the behavior and movement of electrons, affecting its conductivity, insulation, and overall electrical performance.

Entropy (in materials science). p.69   the degree of disorder or randomness in a system, reflecting how the arrangement of atoms or molecules influences the material's thermodynamic stability and behavior.

Grain boundary. p.41   the interface between two crystallites or grains within a polycrystalline material, where the orientation of the crystal lattice changes abruptly, affecting the material's mechanical and thermal properties.

Internal energy. p.66   the total energy contained within a material due to the kinetic and potential energy of its atoms or molecules, influencing its thermal and structural properties.

Interstitial atom. p.38   an atom that occupies a space within the crystal lattice of a material that is not normally occupied by atoms, often leading to changes in the material's properties.

Kinetics (in materials science). Ch.7   the study of the rates and mechanisms of processes such as diffusion, phase transformations, and reactions within materials.

Light-emitting-diode. p.2   a semiconductor device that emits light when an electric current passes through it, commonly used for illumination and display purposes due to its energy efficiency and long lifespan.

Magnetic properties. p.120   The magnetic properties of materials describe how a material responds to an external magnetic field, including its ability to become magnetized, its susceptibility, and its behavior in terms of attraction or repulsion.

Mechanical properties. p.103   The mechanical properties of materials refer to their behavior under various forces and stresses, including attributes like strength, elasticity, hardness, and toughness, which determine their performance and durability in practical applications.

Metallic glass. p.93   a type of amorphous metal that lacks a crystalline structure, resulting in a unique combination of high strength and hardness, along with excellent corrosion resistance.

Microstructure. p.52   the arrangement and organization of their internal features, such as grains, phases, and defects, observable under a microscope, which significantly influences their physical and mechanical properties.

Optical microscope. p.52   a device that uses visible light and a system of lenses to magnify and examine the fine details of small objects or samples, typically at the microscopic scale.

Optical properties. p.114   the optical properties of materials describe how they interact with light, including characteristics such as absorption, reflection, refraction, and transmission, which determine their appearance and functionality in optical applications.

Phase diagram. Ch.6. a graphical representation that shows the stable phases of a material (such as solid, liquid, or gas) under different conditions of temperature, pressure, and sometimes composition.

Quaisicrystal. p.33 a type of solid that exhibits a non-repeating, aperiodic atomic arrangement that lacks the periodicity of traditional crystals but still displays a long-range order and unique symmetry patterns, such as fivefold or sevenfold symmetry.

Quantum computer. p.10, 144 a type of computing device that uses quantum bits (qubits) and principles of quantum mechanics to perform complex calculations at speeds exponentially faster than classical computers for certain types of problems.

Scanning electron microscope. p.54 an instrument that uses a focused beam of electrons to scan the surface of a sample, producing detailed, high-resolution images of its topography and surface features.

Semiconductor. p.130 a material that has electrical conductivity between that of a conductor and an insulator, and whose conductivity can be altered by factors such as temperature, light, or the presence of impurities.

Solid solution. p.67 a homogeneous mixture of two or more elements where the solute atoms are incorporated into the crystal lattice of the solvent material, altering its properties without forming a separate phase.

Substitutional atom. p.36 an atom that replaces an atom of the host material within the crystal lattice, often introducing changes to the material's properties such as strength or electrical conductivity.

Superalloy. p.3 a high-performance alloy designed to withstand extreme temperatures, mechanical stress, and corrosive environments, typically used in aerospace and power generation applications.

Supernova. p.24 a powerful and luminous explosion of a star, marking the end of its life cycle and resulting in the ejection of its outer layers, which can outshine an entire galaxy for a short period.

Thermal properties. p.116 The thermal properties of materials refer to how they conduct, store, and dissipate heat, including characteristics such as thermal conductivity, heat capacity, and thermal expansion, which influence their behavior in various temperature conditions.

Thermodynamics (in materials science). Ch.5 The study of energy transformations and the relationships between heat, work, and the physical properties of materials, which helps predict material behavior under different conditions.

Transistor. p.49   an electronic device that can amplify or switch electronic signals and electrical power, using semiconductor materials to control the flow of current between its terminals.

Transmission electron microscope. p.58   an advanced imaging tool that uses a beam of electrons transmitted through a thin sample to produce high-resolution images of its internal structure at the atomic level.

TTT diagram. p.95   a graph that displays the start and finish times of phase transformations at various constant temperatures, helping predict the microstructures and properties that form during isothermal heat treatments.

Vacancy. p.35   a type of point defect where an atom is missing from its regular position in the crystal lattice, which can affect the material's mechanical and thermal properties.

X-ray diffraction. p.50   a technique used to study the atomic and molecular structure of a material by measuring the pattern of X-rays scattered by its crystal lattice, revealing information about the arrangement and spacing of atoms.

References (Picture Sources)

1. "super" superalloys: Hotter, stronger, for even longer. University of Cambridge. (2008, September 1). https://www.cam.ac.uk/research/news/super-superalloys-hotter-stronger-for-even-longer

2. Administrator. (n.d.). Turbine Blade - Nickel Superalloy. Rolls-Royce UTC. https://www.rrutc.msm.cam.ac.uk/outreach/articles/metals-up-close/turbine-blade-nickel-superalloy

3. H. K. D. H. Bhadeshia, Nickel Based Superalloys. (n.d.). https://www.phase-trans.msm.cam.ac.uk/2003/Superalloys/superalloys.html

4. He, R., Li, M., Han, X., Feng, W., Zhang, H., Xie, H., & Liu, Z. (2023). Experimental Study of As-Cast and Heat-Treated Single-Crystal Ni-Based Superalloy Interface Using TEM. Nanomaterials, 13(3), 608. (The permission of the image use was granted by MDPI's open access policy.) https://doi.org/10.3390/nano13030608

5. Jones, F., & Jones, F. (2023, March 2). Tesla to invest $5bn in Mexican EV gigafactory. Power Technology. (Original image from Tesla) https://www.power-technology.com/news/tesla-mexico-gigafactor/

6. Favreau, J. (2008). Iron Man. © All right reserved by Paramount Pictures.

7. Singer, B. (2006). Superman Returns. © All right reserved by Warner Bros. Pictures.

8. Cameron, J. (1991). Terminator 2: Judgement Day. © All right reserved by TriStar Pictures.

9. Webb, M. (2012). Amazing Spiderman. © All right reserved by Sony Pictures.

10. Type and chemical composition of lithium-ion batteries used in Tesla vehicles. (n.d.). https://servicems.eu/en/news/post/1255-Type-and-chemical-composition-of-lithium-ion-.html

11. Max Roser, Hannah Ritchie – Transistor Count, https://ourworldindata.org/uploads/2020/11/Transistor-Count-over-time.png

12. Staff, N. (2022, June 22). How to Choose a CPU. Newegg Insider. https://www.newegg.com/insider/how-to-choose-a-cpu/

13. Yirka, B. (2020, August 28). Google conducts largest chemical simulation on a quantum computer to date. https://phys.org/news/2020-08-google-largest-chemical-simulation-quantum.html

14. For Educators – GEMS: Gender Equity in Materials Science. (n.d.). https://gems.matse.illinois.edu/educators/

15. "Cosmic Cliffs" in the Carina Nebula (NIRCam Image). (n.d.). Webb. https://webbtelescope.org/contents/media/images/2022/031/01G77PKB 8NKR7S8Z6HBXMYATGJ

16. Crab Nebula (NIRCam and MIRI Image). (n.d.). Webb. https://webbtelescope.org/contents/media/images/2023/137/01HBBMD H12APPEGB8DXVVEP8XA

17. Cameron, J. (2009). Avatar. © All right reserved by 20th Century Fox.

18. Lodovico. (2019, May 20). Crystalline and Amorphous Solids. PhysicsOpenLab. https://physicsopenlab.org/2018/02/13/crystalline-and-amorphous-solids/

19. Zhang X., Schneider R., Müller E., Mee M., Meier S., Gumbsch P., Gerthsen D.1, "Electron microscopic evidence for a tribologically induced phase transformation as the origin of wear in diamond," *Abstract: MS-2-P-1745*. (n.d.). https://www.microscopy.cz/html/1745.html

20. Epionelynx. (2013, February 14). The Bravais Lattices Song. Minerva. https://epionelynx.wordpress.com/2013/02/14/the-bravais-lattices-song/

21. Fisher, I. R., Cheon, K. O., Panchula, A. F., Canfield, P. C., Chernikov, M., Ott, H. R., & Dennis, K. (1999). Magnetic and transport properties of single-grain R−Mg−Zn icosahedral quasicrystals [R=. Physical Review. B, Condensed Matter, 59(1), 308–321. (The permission of the image use was granted by American Physical Society) https://doi.org/10.1103/physrevb.59.308

22. J.W. Evans, The Ames Laboratory, US Department of Energy, Al-Pd-Mn Quasicrystal Surface | NIST. (n.d.). NIST. https://www.nist.gov/image/al-pd-mn-quasicrystal-surface

23. Callister Jr., William D., Rethwisch, David G. (2015). Materials science and engineering: an introduction (9th ed.). New Jersey: John Wiley.

24. File:Microstructure of a stainless steel A961.jpg - Wikimedia Commons. (2019, March 7). https://commons.wikimedia.org/wiki/File:Microstructure_of_a_stainless_s teel_A961.jpg

25. Gupta, R., Mitchell, D., Blanche, J., Harper, S., Tang, W., Pancholi, K., Baines, L., Bucknall, D. G., & Flynn, D. (2021). A Review of Sensing Technologies for Non-Destructive Evaluation of Structural Composite Materials. Journal of Composites Science, 5(12), 319. (The permission of the image use was granted by MDPI's open access policy.) https://doi.org/10.3390/jcs5120319

26. File:XRD diffractometer.svg - Wikimedia Commons. (2020, November 17). https://commons.wikimedia.org/w/index.php?curid=96222184

27. Zhu, X., Yan, H., Zhang, M., & Wei, Q. (2020). A new cubic superhard large-cell carbon allotrope: c-C200. Results in Physics, 19, 103457. (The permission of the image use was granted by Elsevier's open access policy.) https://doi.org/10.1016/j.rinp.2020.103457

28. File:Parts of a light microscope (english) - larger text.png - Wikimedia Commons. (2022, May 2). (Original Source: Mikael Häggström, M.D.) https://commons.wikimedia.org/wiki/File:Parts_of_a_light_microscope_(e nglish)_-_larger_text.png

29. File:Eisen - Meteorit 02.jpg - Wikimedia Commons. (2009, December 20). (Originall Source: H. Zell, Iron meteorite, Octahedrite with Widmannstätten patterns, Gibeon, Southwest Africa, 1836; Staatliches Museum für Naturkunde Karlsruhe, Germany.) https://commons.wikimedia.org/wiki/File:Eisen_-_Meteorit_02.jpg

30. File:As-cast-microstructure-of-Al2O-Mg-sub2sub-Si-composite.jpg - Wikimedia Commons. (2012, April 3). (Original Source: A. Malekan, M. Emamy, J. Rassizadehghani, M. Malekan) https://commons.wikimedia.org/wiki/File:As-cast-microstructure-of-Al2O-Mg-sub2sub-Si-composite.jpg

31. File:Microstructure of a steel powder particle.jpg - Wikimedia Commons. (2013, July 23). (Original Source: Gael Guetard) https://commons.wikimedia.org/wiki/File:Microstructure_of_a_steel_pow der_particle.jpg

32. File:Austempered Ductile Iron (ADI).jpg - Wikimedia Commons. (2012, July 17). (Original Source: Łukasz Boroń) https://commons.wikimedia.org/wiki/File:Austempered_Ductile_Iron_(A DI).jpg

33. File:U.S. Department of Energy - Science - 395 074 001 (33186277393).jpg - Wikimedia Commons. (2016, November 15). https://commons.wikimedia.org/wiki/File:U.S._Department_of_Energy_-_Science_-_395_074_001_(33186277393).jpg

34. Mahadeshwara, M. R. (n.d.). Scanning Electron Microscope – About Tribology. https://www.tribonet.org/wiki/scanning-electron-microscope/

35. USA, J. (2020, January 27). JEOL Introduces New Field Emission SEM with Automated Analytical Intelligence. https://www.jeolusa.com/NEWS-EVENTS/Press-Releases/jeol-introduces-new-field-emission-sem-with-automated-analytical-intelligence

36. http://remf.dartmouth.edu/images/insectPart3SEM/source/3.html

37. File:Ductile Fracture Surface 6061-T6 Al SEM.png - Wikimedia Commons. (2019, December 31). https://commons.wikimedia.org/wiki/File:Ductile_Fracture_Surface_6061-T6_Al_SEM.png

38. File:Nanowires that Emit UV Light (5884303101).jpg - Wikimedia Commons. (2006, May 25). https://commons.wikimedia.org/wiki/File:Nanowires_that_Emit_UV_Light_(5884303101).jpg

39. EDS Mapping. (n.d.). Scanning Electron Microscope Facility. https://semfe.stanford.edu/applications/EDS_Map

40. Science as Art | Galleries of winning images from past MRS meetings. (n.d.). https://www.mrs.org/programs-outreach/science-as-art

41. FEI Titan 80-300 TEM. (n.d.). https://physnano.univie.ac.at/equipment/fei-titan-80-300-tem/

42. File:ODR.Quadri.fig1.png - Wikimedia Commons. (2018, September 1). (Original Source: Sotaro Chiba, José R. Castón, Said A. Ghabrial and Nobuhiro Suzuki from https://talk.ictvonline.org/ictv-reports/ictv_online_report/dsrna-viruses/w/quadriviridae) https://commons.wikimedia.org/wiki/File:ODR.Quadri.fig1.png

43. File:HrtemMg.png - Wikimedia Commons. (2021, November 2). https://commons.wikimedia.org/wiki/File:HrtemMg.png

44. File:Hrtem hematit.png - Wikimedia Commons. (2021, December 9). (Original Source: Mariana Klementová from the Institute of Physics of the Czech Academy of Sciences) https://commons.wikimedia.org/wiki/File:Hrtem_hematit.png

45. Araki, S., Oishi, K., & Terada, Y. (2021). Interface Strengthening of α-Mg/C14–Mg2Ca Eutectic Alloy. Metals, 11(12), 1913. (The permission of the image use was granted by MDPI's open access policy.) https://doi.org/10.3390/met11121913

46. Diffusion. (2024, June 23). Wikipedia. https://en.wikipedia.org/wiki/Diffusion#/media/File:Blausen_0315_Diffusion.png

47. Lumen Learning. (n.d.). *Phase diagrams*. In Chemistry: Atoms First. https://courses.lumenlearning.com/suny-chem-atoms-first/chapter/phase-diagrams-2/

48. R.E. Smallman, A.H.W. Ngan, Chapter 2 - *Phase Diagrams and Alloy Theory*, Modern Physical Metallurgy, 43-91 (2014)

49. "Fe–C Phase Diagram." n.d. MetallurgySite. Retrieved November 27, 2025 (https://metallurgysite.com/learning/matFeCdiagram.php)

50. Dutta, S., Kumar, V., Shukla, A., Mohapatra, N. R., & Ganguly, U. (2017). Leaky Integrate and Fire Neuron by Charge-Discharge Dynamics in Floating-Body MOSFET. Scientific Reports, 7(1) 8257. (The permission of the image use was granted by Scientific Reports's open access policy.) https://doi.org/10.1038/s41598-017-07418-y

51. Qu, Z., Liu, L., Deng, Y., Tao, R., Liu, W., Zheng, Z., & Zhao, M. C. (2022). Relationship between Biodegradation Rate and Grain Size Itself Excluding Other Structural Factors Caused by Alloying Additions and Deformation Processing for Pure Mg. Materials, 15(15), 5295. (The permission of the image use was granted by MDPI's open access policy.) https://doi.org/10.3390/ma15155295

52. Lee, S. W., Huh, M. Y., Fleury, E., & Lee, J. C. (2006). Crystallization-induced plasticity of Cu–Zr containing bulk amorphous alloys. Acta Materialia, 54(2), 349–355. (Permission of the image use was granted by Elsevier.) https://doi.org/10.1016/j.actamat.2005.09.007

53. Metallurgy for Dummies. (n.d.). *Time-temperature-transformation (TTT) diagram.* Metallurgy for Dummies. Retrieved November 26, 2025, from https://www.metallurgyfordummies.com/time-temperature-transformation-ttt-diagram.html

54. Advanced Technical Products. (2023, August 16). *Advanced Technical Products explores various heat treatment techniques.* Advanced Technical Products. https://advancedtechnicalprod.com/industry-news-blog/advanced-technical-products-explores-various-heat-treatment-techniques/

55. x.com. (n.d.). X (Formerly Twitter). https://x.com/RoyNemer/status/1594997199154331651/photo/1

56. За първи път у нас - завод за части за самолети. (n.d.). http://www.karlovo.tv/novinaArch.php?id=56358

57. Woodard, J., & Staff, I. (2019, July 30). Iron Man 3. The Santa Barbara Independent. https://www.independent.com/2013/05/08/iron-man-3/

58. Jackson, P. (2001). Load of Rings. © All right reserved by New Line Cinema.

59. Ashby, M. (2017). Materials Selection in Mechanical Design (5th ed.). Oxford. United Kingdom: Butterworth-Heinemann.

60. A Brief Introduction Of Transmission Towers | Electric Transmission Tower - Utkarsh India. (n.d.). Utkarsh. https://utkarshindia.in/blog/a-brief-introduction-of-transmission-towers

61. Lee, S., Kim, J., Kim, H. T., Im, S., An, S., & Auh, K. H. (2023). Superconductor $Pb_{10-x}Cu_x(PO_4)_6O$ showing levitation at room temperature and atmospheric pressure and mechanism. arXiv (Cornell University). https://doi.org/10.48550/arxiv.2307.12037

62. Pillars of Creation (Hubble and Webb images side by side). (n.d.). Webb. https://webbtelescope.org/contents/media/images/2022/052/01GF44EV 0PPW2BHJS9HMA1AGEK

63. Kosinski, J. (2022). Top Gun: Maverick. Paramount Pictures.

64. Westwood Aerogel. (n.d.). Westwood Aerogel. https://westwoodaerogel.com/

65. 원다라. (2023, August 2). 서울 공중 부양 도시로 바뀐다. "초전도체 밈" 화제. 한국일보. https://www.hankookilbo.com/News/Read/A2023080210540005134

66. Bailey, G. (n.d.). Let's discuss motors in Electric vehicles continued. – European Training Network for the Design and Recycling of Rare-Earth Permanent Magnet Motors and Generators in Hybrid and Full Electric Vehicles (DEMETER). https://etn-demeter.eu/lets-discuss-motors-in-electric-vehicles-continued/

67. 금속, 너 도대체 뭐니? (2020, August 12). Brunch Story. https://brunch.co.kr/@sciforus/182

68. Alpa. (2023, July 11). 알루미나 세라믹스의 열전도도에 대한 분말의 영향. ALPA Powder Technology. https://www.alpapowder.com/ko/127630/

69. Q4 2017 300 mm Silicon Wafer Pricing to Increase 20% YoY in DRAM-like Squeeze. (n.d.). TechPowerUp. https://www.techpowerup.com/238785/q4-2017-300-mm-silicon-wafer-pricing-to-increase-20-yoy-in-dram-like-squeeze

70. Lamborghini car features carbon fiber structure. (2024, May 21). Reinforced Plastics. https://www.reinforcedplastics.com/content/products/lamborghini-car-features-carbon-fiber-structure/

71. NASA Invites Public to Take Flight With Ingenuity Mars Helicopter - NASA. (n.d.). NASA. https://www.nasa.gov/news-release/nasa-invites-public-to-take-flight-with-ingenuity-mars-helicopter/

72. File:Bioenergy (4101037876).png - Wikimedia Commons. (2009, November 13). (Original Source: Simulation: Jeremy Smith, University of Tennessee and ORNL. Visualization: Jamison Daniel, ORNL). https://commons.wikimedia.org/wiki/File:Bioenergy_(4101037876).png

73. Langer, R. (2024, February 20). How Nanotech Powers Precision Medicine. Scientific American. (Getty Image) https://www.scientificamerican.com/blog/observations/how-nanotech-powers-precision-medicine/

74. Admin_Indivenire. (2019, September 16). Biomaterials - R&D - Outsourced Research and Development - Indivenire. https://indiveni.re/biomaterials/

75. Marine Bacteria Can Degrade Plastic Pollution. (2023, October 16). Applied Sciences From Technology Networks. (Naja Bertolt Jensen / Unsplash.) https://www.technologynetworks.com/applied-sciences/news/marine-bacteria-can-degrade-plastic-pollution-379920

76. India, A. T. (2023, October 16). AI in India's Renewable Energy Sector: Propelling Sustainability. aitrendsindia.com. https://aitrendsindia.com/artificial-intelligence/ai-in-indias-renewable-energy-sector-propelling-sustainability/

77. Additive Manufacturing 3D Printing. (2015, November 12). PureAire Monitoring Systems Oxygen Monitor. https://www.pureairemonitoring.com/additive-manufacturing-3d-printing-the-growth-progress-and-need-for-safety-monitors/

78. Technology | IBM Quantum Computing. (n.d.). https://www.ibm.com/quantum/technology

79. Veritasium. (2014, August 13). *Explained: 5 Fun Physics Phenomena* [Video]. YouTube. https://youtu.be/jIMihpDmBpY

Author, Character Illustration: Seok-Woo Lee

- Bachelor's degree, Materials Science and Engineering, Korea University, South Korea (1997~2004)
- Master's degree, Materials Science and Engineering, Korea University, South Korea (2004~2006)
- PhD, Materials Science and Engineering, Stanford University, USA (2006~2011)
- Postdoc, Applied Physics and Materials Science, California Institute of Technology (2011~2014)
- Associate professor, Materials Science and Engineering, University of Connecticut (2014~present)

Author: Wyeth Haddock

- Bachelor's degree, Materials Science and Engineering, University of Connecticut (2022~present)

www.ingramcontent.com/pod-product-compliance
Lightning Source LLC
Chambersburg PA
CBHW070923130626
46555CB00001B/260